MOTILE
MUSCLE AND CELL
MODELS

STUDIES IN SOVIET SCIENCE

MOTILE MUSCLE AND CELL MODELS

N. I. ARRONET
Institute of Cytology
Academy of Sciences of the USSR
Leningrad, USSR

**Translated from Russian by
Basil Haigh**

 CONSULTANTS BUREAU • NEW YORK–LONDON • 1973

Nikolai Iosifovich Arronet was born in 1931 in Leningrad. In 1954 he was graduated from the Soil-Biology Department of the Central Asian State University (Tashkent) where he specialized in animal physiology and biochemistry. Since 1955 he has been working in Leningrad under the supervision of Prof. V. Ya. Aleksandrov, first at the Zoological Institute and then at the Institute of Cytology, Academy of Sciences of the USSR; he is presently a junior scientific co-worker at the latter institute. In 1967 he received the degree of Candidate of Biological Sciences. Intracellular mechanisms of the regulation of motor activity of the cell and their evolution are his scientific interests at present.

The original Russian text, published by Nauka Press in Leningrad in 1971, has been corrected by the author for the present edition. This translation is published under an agreement with Mezhdunarodnaya Kniga, the Soviet book export agency.

MYSHECHNYE I KLETOCHNYE SOKRATITEL'NYE (DVIGATEL'NYE) MODELI
N. I. Arronet

МЫШЕЧНЫЕ И КЛЕТОЧНЫЕ СОКРАТИТЕЛЬНЫЕ (ДВИГАТЕЛЬНЫЕ) МОДЕЛИ
Николай Иосифович Арронет

Library of Congress Catalog Card Number 72-88884
ISBN 0-306-10877-1

© 1973 Consultants Bureau, New York
A Division of Plenum Publishing Corporation
227 West 17th Street, New York, N.Y. 10011

United Kingdom edition published by Consultants Bureau, London
A Division of Plenum Publishing Company, Ltd.
Davis House (4th Floor), 8 Scrubs Lane, Harlesden, London, NW10 6SE, England

All rights reserved

No part of this publication may be reproduced in any form without written permission from the publisher

Printed in the United States of America

Foreword

One of the early ways by which man learns about the surrounding world is by taking to pieces an object which attracts his attention. This method, which is widely used by children, is resorted to once again by the adult who wishes to study natural phenomena. The method of dismantling a complete object into its parts so that it can be studied has found its widest and most successful application in biology and, in particular, in the study of living cells. The cytologist studying a cell will usually have to examine its fragments, which may be dead or semiliving. In sections of killed cells, with the aid of the light or electron microscope he tries to obtain some idea of the structure of the living cell. As he investigates fractions isolated from a tissue homogenate, or substances isolated from individual cells, he tries to discover the biochemical functions of the cell organoids and their components. In every case the study of degradation products of the cell (depending on the degree of degradation these may be isolated nuclei, nucleoli, nucleolar ribosomes, ribosomal proteins and RNA, mitochondria, fragments of mitochondrial membranes, and so on) makes its own contribution to knowledge, and the information obtained on different objects is mutually complementary.

Among all the cell fragments obtained experimentally which can make a great contribution to our knowledge of cell structure, an important place is occupied by contractile cell models. These are the remains of cells from which components which have nothing to do with the contractile function have in some way or other been removed or destroyed; the contractile system is left structurally and functionally intact as far as possible. Such cell models afford fresh opportunities for the study of the contractile system from many different aspects: by comparison with intact living cells and also by comparison with their degradation products — the contractile proteins.

This book gives an account of the history of development of methods used to prepare cell models and the distinctive features of models obtained from different objects, it outlines the sphere of their application to the solutions of problems in cell physiology, and it gives details of the methods used to prepare a number of cell models. The book mentions many of the interesting and important results already obtained by the use of cell models. However, the main purpose of N. I. Arronet's book, it seems to me, is to encourage the use of the method as one which could prove useful for many researchers working on cytological and general physiological problems. Cell models are also of considerable interest from the teaching point of view, especially in connection with the energetics of cell motility. They have been used with great success in the teaching of cytology in the Department of Cytology and Histology of Moscow University.

In work with cell models, however, the following fact must not be overlooked. Sometimes the investigator becomes so fascinated by the breakdown products that he tends to forget the whole. In that case the means becomes the end and the whole strategy of research deviates from its course. This must be avoided in work with cell models which are themselves so interesting that it is easy to lose sight of the living cell as a whole.

V. Ya. Aleksandrov

Contents

CHAPTER I. The Concept of Cell Model. The General Characteristics of Models, Principles of Their Preparation, and Their Sphere of Applicability........... 1
CHAPTER II. Properties of Various Motile Models.............................. 21
  1. Contraction evoked by ATP................ 21
    Muscle models...................... 21
    Models of contraction of nonmuscle cells and their organoids....................... 54
  2. Lengthening induced by ATP............... 72
  3. Movements composed of a single combination of ATP-induced contraction and relaxation....... 74
    Movements in anaphase of mitosis.......... 74
    Movements in telophase of mitosis......... 81
  4. Contraction induced by calcium ions and lengthening induced by ATP..................... 90
  5. Lengthening induced by calcium ions and inhibited by ATP............................. 95
  6. ATP and cytoplasmic streams in plant cells.... 96
  7. Rhythmic movement of flagella and cilia....... 99
CHAPTER III. The Use of Motile Models to Study Problems of a Nonmechanochemical Nature........................ 115
CHAPTER IV. Methods of Obtaining Motile Models.............................. 131
  Preliminary remarks...................... 131

Actomyosin threads	132
Actomyosin threads from cross-striated skeletal muscles	132
Actomyosin threads from smooth muscles	134
Models of the muscle fiber	135
Models of cross-striated skeletal muscle fibers	135
Preparation of model fibers of the flying muscles of insects	137
Models of smooth-muscle fibers.	138
Preparation of glycerinated models of muscle fibers from the ventricular myocardium	140
Isolated myofibrils	141
The muscle fiber without its sarcolemma.	143
Muscle fiber with disrupted coupling in the sarcotubular system	144
Preparation of isolated protofibrils	144
Preparation of isolated sarcolemma suspension	145
Preparation of relaxing granules from skeletal muscles	145
Cell (interphase) models becoming rounded in the presence of ATP	146
Anaphase models of fibroblasts	147
Models of cleavage in the sea urchin egg	148
Telephase models of fibroblasts	149
Models of mitochondria.	150
Models of chloroplasts	151
Models of locust and frog spermatozoa	151
Models of mammalian spermatozoa.	152
Models of spermatozoa of certain marine invertebrates	153
Models of ciliated epithelium	153
Models of the Vorticella stalk	155
Model cilia from infusorians	156
Models of flagella of unicellular algae	159
Models of flagella from protozoans	160
Models of amebas	160
Protoplasmic streaming models in the marine alga <u>Acetabularia calyculus</u>	161
Models of the myxomycete plasmodium	162

The use of models for practical work with students ..	162
ATP and muscle contraction	162
ATP and ciliary movement	163
NOTES ADDED IN PROOF	167
BIBLIOGRAPHY	177
Surveys	177
Experimental papers	180

Chapter I

The Concept of Cell Model. The General Characteristics of Models, Principles of Their Preparation, and Their Sphere of Applicability

Although mechanical activity is manifested so clearly by the function of a muscle, it is by no means confined to muscle tissue. Since in everyday life, at every step (metaphorically as well as literally) we come face to face with this form of protoplasmic motor activity, we have become accustomed to regard the contraction and relaxation of a muscle as the pinnacle of biomechanical activity. In reality, however, contractility of the muscle fiber is only one such pinnacle, and possibly not the highest. There are many other forms of contraction, lengthening, and movement of cells, their organoids, and components — some of them highly sophisticated and widespread. Conformational changes in enzymes in the course of their function, the enzymic transport of materials through membranes, the contraction of the membranes of mitochondria and chloroplasts, the movement of ribosomes along the molecule of messenger RNA, the contraction of the phage sheath after attachment of the phage particle to the cell, the movement of bacterial flagella, cytoplastic streams, the rounding of tissue cells in response to excitation or injury, the many movements taking place during mitosis (rounding of the cell in prophase, movement of the chromosomes to opposite poles simultaneously with lengthening of the cell in anaphase, division of the cell by the formation of a central constriction into two daughter cells in telephase) and in myosis, the contraction and relaxation of the myonemes of protozoans, the ameboid movement of unicellular organisms and tissue cells, the movement of the flagella of algae, spermatozoa, and choanocytes, the oscillation of undulating membranes, of the flagella, cilia, and their derivatives in protozoans, and the cilia of ciliated epithelial cells, the contraction and relaxation of the muscle fiber — all these are different manifestations of that protoplasmic mechanical activity which is an intrinsic property of living matter. They were all created and have diverged and acquired their present form in the course of evolution, and it would now be difficult even to arrange them in order of increasing complexity or value.

The old idea that the various types of motor function of cells are based on related or even identical mechanisms (Kühne, 1864,* etc., etc.) was looked upon as something which was almost obvious, but no evidence could be found to support it until the mechanism of each form of movement had been at least approximately explained. The way in which the contractile system of muscle functions was the first problem on which some light has been shed.

The mechanism of muscular contraction has been investigated by analytical methods, by isolating the individual components of the contractile system from the muscle and studying them separately. The crucial moment in this activity was the discovery of the ATPase properties of (acto)myosin in 1939 by V. A. Éngel'-gardt and M. N. Lyubimova (Lyubimova and Éngel'gardt, 1939; Éngel'gardt and Lyubimova, 1939).

The study of isolated contractile proteins could solve a number of problems of the utmost importance: the substrate of the mechanochemical reaction, the kinetics of that reaction, the conditions which activate it or inhibit it, the presence or absence of conformational changes in molecules of contractile proteins during their interaction with ATP, and so on. These are all essential approaches to the elucidation of the mechanism of muscle contraction; however, many essential problems concerned with the mechanism of muscle contraction cannot be solved by the study of protein in solution. The reasons for this are as follows: the solution of a protein has no structural conformation, the molecules of its contractile proteins are not arranged in a definite order relative to

---

*Years of publication of surveys are underlined, years of publication of experimental results are not. Lists of both are given separately at the end of the book.

one another as they are in the myofibril; the concentration of a protein in solution cannot be as high as it is in the myofibril; in the myofibril the contractile proteins are in a complex and specific chemical environment which cannot be reproduced in solution. In other words, contractile proteins in solution are in a state which differs considerably from that in which they exist in the muscle fiber. As a result it is impossible to investigate the biophysical problems of contraction, its mechanics, its thermodynamics, and its kinetics in solutions of contractile proteins. The need thus arose for preparations with contractile components in an orderly arrangement. Ideally, this orderliness would correspond to the structure found in the normal muscle fiber. However, the living fiber usually cannot be used for this purpose because the study of its contractile skeleton is complicated by other systems: those which keep the fiber alive, generate high-energy compounds, and conduct excitation and transmit it to contractile structures, i.e., systems not directly participating in the mechanochemical act of contraction. Selective permeability of the living fiber, preventing the action of certain reagents on the contractile system, create serious difficulties for the investigator. When he studies the effect of a substance on the living muscle fiber, just as on any other cell, the physiologist is primarily — and often entirely — studying the action of this substance on the cell membrane and not on the internal regions of the fiber where the contractile system is situated. Adenosine triphosphate, which by its action on the living fiber can cause it to contract, in fact acts in this situation as a nonspecific stimulus depolarizing the membrane, and not as the source of energy and the substrate for the mechanochemical reaction.

Two possibilities remain: first, to "simplify" the fiber by removing from it those systems which maintain and regulate contractility but interfere with the experiment, and second, to increase the complexity of a solution of contractile proteins by forming preparations with a regular molecular arrangement from it. Both these possibilities have been used to study the mechanism of muscular contractions and, as a result, a wide range of preparations simulating contraction of the muscle fiber and reproducing more or less closely the actual conditions pertaining in the fiber, have been suggested. These preparations form a series with gradations of orderliness of their structures, starting with the protein solution with its total lack or organization (or with artifactual organization) of the molecules of contractile protein and ending with the living muscle fiber.

Preparations, obtained from cells, capable of exhibiting the mechanochemical effect under suitable conditions but deprived of their selective permeability (and, consequently, incapable of excitation or of conduction and transmission of excitation), of their synthetic activity, and of their systems of movement control and coordination have been given the name "contractile models." The definition "contractile" is also applied to models in which the movement produced by them is not a contraction but, quite the reverse, a lengthening of the structures. Strictly speaking they should be described by the more general term of "motile models." Models reproducing the elements of muscle movements are called muscle models. The analogous models prepared from cells which are not muscle cells, or from their organoids, but which perform their own characteristic movements, are called cell models. They are nonliving systems but they have the ability to perform a mechanochemical reaction. The difference between the state of the protein components of a model and the state of an enzyme in vitro is that models still preserve a high degree of structural orderliness such as characterized the mechanically active system in vivo. Modern biochemistry, however, has also commenced the study of noncontractile enzymes isolated in a structurally organized form.

The simplest way of orienting long protein molecules, such as the molecules of myosin and actomyosin, is by creating a flow of protein solution. In such a flow the molecules lie with their long axis in the direction of movement. This is shown by the appearance of birefringence in a flowing solution of such a protein (Needham et al., 1941).

The next step in recreating the structural relationships approximating to those in the myofibril was to prepare threads from isolated actomyosin in which the molecules are arranged relatively uniformly. Threads of this type were obtained from (acto)myosin by Weber (1934) by extruding a solution of this protein form a capillary tube into water. On entering the water, the stream of protein congeals and is converted into a gel. A thin thread is formed. Since it is a flowing solution which congeals, i.e., a solution with regularly oriented molecules, the molecules in the thread are also oriented along its length.

The character of orientation of the molecules in the thread is demonstrated by the results of x-ray structural analysis and by their inherent birefringence (Needham, 1950). Éngel'gardt, Lyubimova, and Meitina (1941), who considered that (acto)myosin threads provide some degree of approximation to the myofibril, investigated their mechanical properties and the effect of various chemical agents on them. They found that ATP sharply increases the plasticity of the threads, and if they are under tension they stretch considerably. This investigation essentially marks the beginning of mechanochemistry. Almost simultaneously Szent-Györgyi (1941-1942) showed that actomyosin threads in a free state, not under tension, contract if placed in ATP solution. The mechanochemical activity of the threads is directly connected with the ATPase activity of their contractile protein, and is due to it: during storage the threads lose both types of their activity concurrently, and the same inhibitors depress both types of activity equally.

However, the degree of orientation of the protein molecules in the threads is much lower than in the myofibril. They lack the orderly disposition of the myosin and actin elements relative to each other which characterizes the myofibril, in which they are responsible for the picture of cross-striation.

Again, the protein concentration in the actomyosin thread is only one-quarter that in the myofibril. One result of the incomplete orderliness of arrangement of the molecules in the actomyosin thread is that the character of its contraction under the influence of ATP differs from the character of contraction of the muscle fiber. Not only does the thread shorten, but it also becomes thinner (isodimensional contraction), whereas a fiber which contracts, on the contrary, increases its thickness (anisodimensional contraction). During contraction the thread expels and loses a large quantity of water, which does not take place during contraction of the muscle fiber.

Moss and co-workers (Moss et al., 1935, cited by Éngel'gardt and Lyubimova, 1942) obtained monomolecular films of myosin by pouring the protein on the surface of a solution of lactate. The protein molecules in films of this type are arranged in very regular order. The workers who obtained this preparation did not use it for mechanochemical investigations. The proposal that this should be done was made by Éngel'gardt and Lyubimova (1942); they pointed out that the two-dimensional structure of the film preparation, which is not typical of the myofibril, is a drawback. Nevertheless, film preparations of contractile proteins were used later, when Hayashi (1951, 1952) suggested the method of preparing actomyosin threads by compression from monomolecular films. The degree of orientation of the molecules in these preparations is far higher than in films obtained by extruding protein from a capillary tube. Because of this, threads obtained from films contract in the presence of ATP only in the direction of their long axis and do not become thinner.

In the cases mentioned above (contractile protein in a stream, actomyosin threads) the contractile protein or protein complex is extracted from muscles and is then given the molecular orientation which simulates the molecular structure of this protein in the living fiber. Actomyosin which has been given an orderly structure in this way, is separated from all other components of the cell which surround it and provide for it in vivo, and if placed in a medium containing ATP as the source of energy, will perform mechanical work.

However, as was mentioned above, it is possible to proceed by a different way, by removing noncontractile components from the fiber (or cell or organoid), and so leaving the actual contractile system itself exposed but untouched in its proper place and accessible for study. The contractile apparatus of a fiber or cell killed and extracted in this way can be used to create the appropriate experimental conditions for mechanical work of the sort which it performs in the living fiber, cell, or organoid.

The first contractile model of this type was obtained by Varga in Szent-Györgyi's laboratory (Varga, 1946; Szent-Györgyi, 1949) from muscle fibers killed by glycerol extraction. Later, by making use of the principle of obtaining the muscle model and as a result of the experience gained by working with it, models were successfully obtained from a wide variety of cells whose contractile system is able to do mechanical work of a different character from muscular contraction.

In 1953-1954, Hoffmann-Berling made models of connective-tissue cells which reproduce the contractile rounding of the cell bodies as well as the various intracellular and cellular movements taking place during mitosis. In 1955 the same worker obtained models of the tails of spermatozoa. This was the first model which performed not merely a single isolated contraction of lengthening, but a complete work cycle repeated over and over again, i.e., a rhythmic movement. In 1956, Aleksandrov and Arronet made similar models of the cilia of the ciliated epithelium of a number of vertebrates. Model cilia flicker in the presence of ATP and Mg^{++}. Soon after this Hoffmann-Berling, followed by the present writer, also obtained models of ciliary movement of infusorians. Levine (1956) obtained and Hoffmann-Berling (1958) investigated models of the Vorticella stalk, and showed that the mechanism of contraction differs radically from that of muscular contraction. Later other contractile models were suggested: amebas, the plasmodium of a myxomycete, mitochondria, chloroplasts, rhythmically moving models of flagella of lower plants, and a model of the movement of the cytoplasm of a plant cell. Features of similarity and difference between the various forms of cell movement were revealed by means of these models, and the way in which these differences are attributable to the properties of the motor system were studued. New methods of preparing models also were developed.

The glycerinated muscle fiber will serve as an example when describing the general characteristics of contractile models and the principles governing their preparation.

If a muscle is placed in a 50% solution of glycerol in water or weak salt solution and kept in this solution in the cold for, let us say, 24 h the muscle will die. If, however, a muscle killed in this way is split into single fibers or bundles of a few fibers each, washed to remove glycerol, and then placed in neutral solution containing ATP in physiological concentration, i.e., close to its normal concentration in the living fiber, the dead fiber will contract. It cannot stretch to its initial length spontaneously. However, there are various ways by which such a fiber can be made to relax. After relaxation of the model fiber, it can again be made to contract by placing it in the same solution of ATP.

The contractile system in the fiber model remains in the same form as in the living fiber. At the same time, the other properties of the living fiber — its selective permeability, excitability, ability to accumulate energy supplied to it during dissimilatory processes — are not present in the model fiber. This model thus provides the investigator who uses it with the means of studying a mechanochemical reaction taking place normally during contraction of a muscle, yet in isolation from other processes and uncomplicated by them. The most important feature is that the contractile structures of the preparation are accessible to treatment by the various reagents used to analyze the contraction mechanism.

The advantage of the fiber model over actomyosin threads is that the state of the contractile system in the model is far closer to its state in the living fiber. In the model its components retain their mutual arrangement. Electron-microscopic investigations (Hanson, 1952; Hanson and Huxley, 1955; Abbott and Chaplain, 1966; Jagendorf-Elfvin, 1967; Sjöstrand and Jagendorf-Elfvin, 1967; Komissarchik and Shapiro, 1969, 1971; Ashhurst, 1969) have shown that the submicroscopic morphology of the sarcomere remains almost unchanged in model fibers or myofibrils (Figs. 1-4).

The process of preparation of models from fibers or isolated myofibrils does not change the mutual arrangement of the contractile protein molecules in the structures formed by them, or the protein concentration in those structures. The model fiber retains its sarcolemma. It has lost its selective permeability, but it maintains the integrity of the fiber and preserves the normal structural relationships between the contractile elements during contraction of the model.

On the other hand, there is evidence of incomplete isolation of the contractile system in the glycerinated fiber. Besides proteins of the actomyosin complex, it still contains many other functional proteins. Actomyosin accounts for about 50% of all proteins in the living muscle fiber. When a model is prepared from the fiber about 20% of the proteins other than actomyosin is extracted. Consequently, 30% of all the noncontractile proteins still remains in the model. In particular, the cytochrome system can still function in the fiber model (Wilson et al., 1959). Enzyme systems remaining in the fiber are evidently incomplete, the enzymes composing them no longer function together, and the systems themselves work independently of each other and of the contractile apparatus, but for this very reason they do not interfere with the investigation of the working of the isolated contractile system. The technique of preparing fiber models has now reached a high level

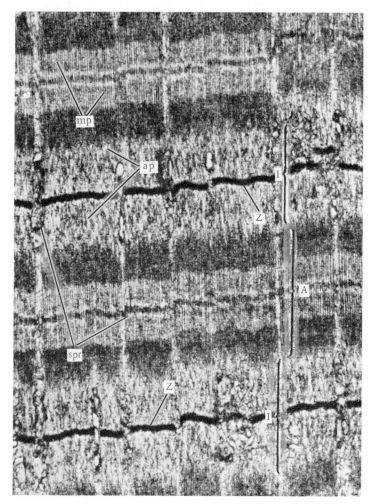

Fig. 1. Glycerinated fiber of rabbit psoas muscle (from Sjöstrand and Jagendorf-Elfvin, 1967). Legend: A- and I-disks and Z-membranes are identified; ap) actin protofibrils, mp) myosin protofibrils, spr) elements of sarcoplasmic reticulum.

of perfection: the virtually total extraction of noncontractile proteins from fibers while leaving the properties of their contractile skeleton intact has been successfully achieved (Abbott and Chaplain, 1966). The misgivings mentioned above are completely eliminated with respect to these detergent-treated models.

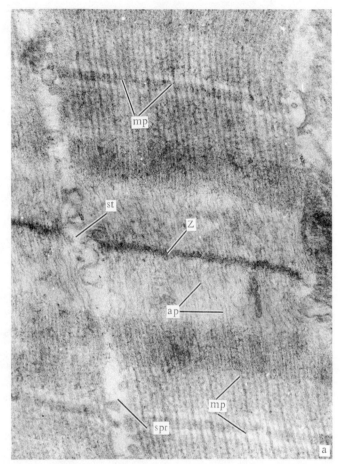

Fig. 2. Model and normal fibers of sartorius muscle of Rana temgraph of Komissarchik and Shapiro); (b) normal fiber (original phofibrils, mp) myosin protofibrils, st) elements of sarcotubular system, branes are identified.

The use of models of muscle fibers is a legitimate, convenient, and useful method of analytical study of muscle contractility. However, the experimental use of fiber models does not preclude the use of other preparations simulating muscle contraction. It is often desirable to plan an investigation by working with several preparations which differ in the degree if isolation of the contractile system or its components from the muscle. Several preparations of this type exist: solutions of the contractile proteins them-

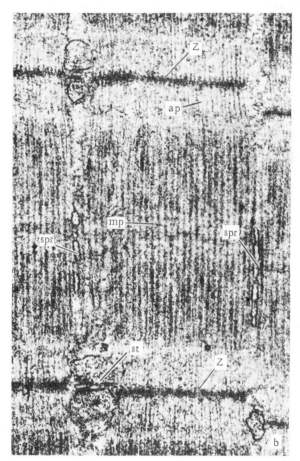

poraria, 45,000×: glycerinated fiber (original photo-
tograph of Krolenko and Adamyan). ap) actin proto-
spr) elements of sarcoplasmic reticulum; Z-mem-

selves; a solution of the protein complex (actomyosin); a flowing solution of actomyosin; actomyosin gel; a suspension of actin and (or) myosin filaments, natural or synthetic; actomyosin threads, conventional or made from film; isolated (glycerinated or not) myofibrils; a model muscle fiber. This provides a gradation from the simplest and most isolated system to that nearest to the living fiber. This series could in fact be completed with the living fiber. Each of these preparations has its own particular advantages. One

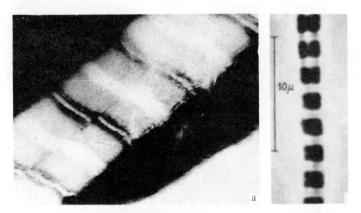

Fig. 3. Model myofibrils from rabbit psoas muscle: (a) myofibril isolated by collagenase method and extracted with 0.078 M borate buffer, pH 7.1, 18,000× (from Perry, 1955); (b) myofibril obtained by mechanical homogenization of glycerinated fibers, phase-contrast microscope (from Hanson and Huxley, 1955).

preparation may be suitable and sufficient for the solution of certain problems. Often, however, an investigation conducted consecutively or simultaneously on several preparations will yield successful results because the advantages of each are combined and their disadvantages are neutralized.

The principle by which models are prepared must now be examined. It is to remove from the cell everything which would complicate the course of the elementary mechanochemical process, without disturbing the structures directly responsible for this process; at the same time access to the contractile systems must be provided for the substances whose action is to be studied. This requires the removal of the obstacle of selective permeability created by the cell membranes.

Consequently, cells to be converted into models must be treated by solutions which rupture or loosen (i.e., partially rupture) the structure of their membranes, after which the globular proteins and compounds of lower molecular weight must be extracted from the cell. It is essential that as a result of the extraction the ATPase of the membranes should be inactivated. The extraction must not disturb the native state of the contractile proteins.

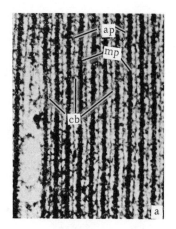 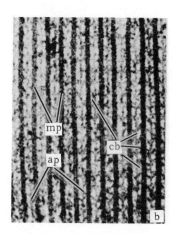

Fig. 4. Glycerinated myofibrils of dorsalis longus muscle of Lethocerus maximus: (a) in state of relaxation arising in medium with ATP and Mg^{++}; (b) during contracture developing in absence of ATP (from Reedy et al., 1965). ap) actin protofibrils, mp) myosin protofibrils, cb) cross bridges, 180,000×.

The extracting solution is usually an aqueous solution of glycerol with salts of low ionic strength (usually about 0.15). For some objects, however, specific substances to dissolve lipids or produce lysis of membranes are added to the extracting solution. These substances include various detergents — saponin, digitonin, Triton X-100, and emulsifying agents (substances of the Tween type, etc.). These substances must not be so rough in their action that they produce complete lysis of the structures necessary to perform the motor acts. Extraction of the cells by detergent solutions is therefore usually a short procedure and the detergent concentration in the solution is low. Excellent results (as regards the completeness of extraction of noncontractile proteins and preservation of the contractile) have been obtained by successive treatment of muscle fibers first with an aqueous solution of glycerol and next with solutions containing detergent (Abbott and Chaplain, 1966).

A method of preparing models in which the membranes are destroyed by tryptic digestion under mild conditions has also been suggested (Schick and Hass, 1949). Perry (1951) has developed a method of obtaining models of myofibrils by digesting the sarcolemma with collagenase, an enzyme which has no action on proteins of the actomyosin complex.

The membranes can be completely or partially destroyed in two ways: either by dissolving their lipids (as has already been stated, this is done by means of detergents) or by removal of their proteins. Enzymic digestion of proteins has not yet been much used in the preparation of models, and neither the combined procedure of partial digestion of proteins and solution of the lipids nor the enzymic hydrolysis of the lipids has yet been tested at all. The possibility is not ruled out that these methods could be useful in the preparation of models from various objects.

Seravin (1963) suggested destroying the membranes by intensively heating the cells and then extracting them with salt solution; this method was found suitable for preparing models of the <u>Vorticella</u> myoneme.

Membranes can also be destroyed by rapid freezing and thawing of the objects (Borbire and Szent-Györgyi, 1949), after which extraction can be carried out (Seravin, 1963). At low temperatures, hydrophobic interactions are weakened, and this could be the basis of this last method of membrane destruction, since hydrophobic interactions undoubtedly play an important role in the stability of their structure. Globular proteins are extracted from cells with partially destroyed membranes by KCl and NaCl contained in the extracting solutions.

The following methods ensure preservation of the mechanically active proteins in the intact state: extraction in the cold (0 to +4°C); keeping in a solution of neutral pH; addition of substances containing SH-group (cysteine, thioglycollate) in order to bind ions of the heavy metals, which can always be present as impurities, and thus to protect the SH-groups of the contractile proteins, which are thiol enzymes; the addition of glycerol as a stabilizer to the extracting solution to prevent denaturation of the proteins; the addition of EGTA (ethyleneglycol-bis-(β-aminoethyl ether)-N,N'-tetraacetic acid) or of EDTA (versene) to bind calcium ions, the presence of which could cause contraction or, conversely, swelling and rupture of the model in the course of its preparation; the addition of polyvinylpyrrolidone or agar to prevent colloidal swelling of the models.

Every object has its optimal extraction time. The shortest possible extraction time is defined as the time during which adequate destruction of the cell membranes takes place so that the selective permeability of the cells is destroyed and their "relaxation system" ceases to function. In the muscle fiber this system is composed of structures of the sarcoplasmic reticulum, while in other cells it is evidently formed by elements of the smooth endoplasmic reticulum. The function of the relaxation system is essentially to pick up calcium ions from the cytoplasm and accumulate them in the tubules and spaces of the endoplasmic (or sarcoplasmic in the case of muscles) reticulum. Since the mechanochemical reaction of contractile proteins with ATP cannot take place in the absence of calcium, the relaxation system keeps the contractile apparatus in a relaxed state. Physiological excitation causes calcium ions to pass into the cytoplasm and to come into contact with the contractile elements, leading ultimately to contraction. When models of fibers or cells are prepared the relaxation system must be destroyed so that it will not prevent the mechanochemical reaction in which the participants are contractile proteins, ATP, Ca^{++}, and Mg^{++}. The ATP and Mg^{++} added from outside then come into contact with the motile system, a mechanochemical reaction takes place and the model moves. Sufficient Ca^{++} is contained in the ATP preparations (Seidel and Gergely, 1963).

From what has been said it will be clear why the preparation of models is inevitably preceded by the destruction of the membranes forming the relaxation system of the fiber or cell.

It must be borne in mind that preparations extracted for insufficiently long periods of time can work as models provided that there is an excess of calcium ions in the medium. This will easily be understood because the relaxation system, although still functioning, is flooded and saturated with calcium ions, so that it cannot accumulate all those which are present in excess. As a result the calcium ions come into direct contact with the contractile structures in concentrations sufficiently high to enable the mechanochemical reaction to take place. Only one additional factor is required in this case – the addition of ATP.

Too long extraction may lead to gradual denaturation of the contractile proteins,* to loss of the functional capacity of the models, and even to their destruction.

---

*A steady decrease in the resistance of models of the ciliated epithelium of the frog palate to heat was observed by Aleksandrov and Arronet (1956) with an increase in the time of their keeping in the extracting solution at 0°C.

The optimal extraction time, depending on the object and the composition of the extracting solution, varies from 30 min to 48 h. However, models can be kept for long periods in special preserving solutions. The preserving solution contains a high concentration of glycerol (for example, 95%), and models are kept in it at a low temperature (between -17 and -20°C). Under these conditions models retain their motility for many months under the influence of the appropriate substances — as a rule, ATP, Mg^{++}, and Ca^{++}.

The glycerol in the preserving solution acts as an antifreeze and also as a substance specifically protecting the proteins against spontaneous denaturation. The antidenaturation effect on model proteins was shown by special experiments in which it increased the resistance of cell models to the action of denaturing agents such as heat and a high hydrostatic pressure (Arronet, 1964a). Further evidence is given by the following fact. When kept in distilled water for 5 h, threads of rabbit actomyosin are denatured and lose their contractility. In 80-90% glycerol these threads remain capable of contracting for weeks or even months (Portzehl, 1951).

Before an experiment is carried out on a model in the course of which its motility will be evoked and observed ("reactivation of the model"), it must first be washed to remove the extracting or preserving solution. When this is done, the glycerol or the lytic agents contained in the extracting solution are removed from the model. The washing solution is a neutral salt solution. The models are usually treated with several portions of washing solution at close to zero temperature for 10-20 min, leading to a stepwise decrease in the concentration of the substance being removed.

The washed model is placed in reactivating solution for observations on its work. Except in special cases, this solution contains ATP or another nucleoside triphosphate. It must also contain Mg^{++} ions to activate the contractile ATPase, and the pH of the solution must be close to neutral. The ionic strength for most objects is 0.12-0.15; for some it must be reduced as much as possible: in that case the ionic strength is created entirely by ATP salts, magnesium, and substances added to buffer the solution; if the relaxation factor has to be suppressed, the ionic strength is increased to 0.22. When certain objects are used, the reactivating solution must contain $CaCl_2$.

THE CONCEPT OF CELL MODEL   17

If polyvinylpyrrolidone or agar is added to the extracting solution, the same substance must be added to the washing and reactivating solutions.

The model must not be very thick, a restriction which applies mainly to muscle. Fairly thick bundles of muscle fibers can be extracted, and the whole muscle can be immersed in the extracting solution in the case of the sartorius and certain other muscles of the frog. However, thick preparations such as these cannot be made to contract in the reacting solution. The reason is that the surface layers of the fibers of a model muscle hydrolyze ATP so intensively that it cannot penetrate into the inner layers. As a result, only the outer one or two layers of fibers will contract, and the force of contraction which they develop is too small to cause the whole muscle or bundle of fibers to contract. That is why, in work with models of the muscle fiber it is preferable to use single fibers or bundles of not more than two or three fibers.

Extraction is best carried out not with single fibers, but with bundles of a certain thickness. During dissection of living fibers the slightest mechanical injury to them causes spreading Zenker's necrosis, which completely inactivates the contractile system. After extraction of the bundle, it must be separated into single fibers. It is convenient and safe to dissect glycerinated fibers, however, and this can be done before the model muscle has been washed to remove the glycerol extracting solution: the models do not undergo Zenker's necrosis.

One other method aimed at securing diffusion of ATP into the interior of the cell model must be described. The model is placed in reactivating solution at 0°C, left in it at that temperature for 5-15 min, and then transferred into the same solution but at room temperature or, if the experiment requires it, at a higher temperature. At 0°C the components of the reactivating solution merely diffuse into the model, but virtually no mechanochemical reaction takes place. The reason is that, if the temperature is lowered to 0°C, the rate of diffusion is reduced much less than the rate of ATP hydrolysis. As a result, the model becomes permeable throughout with reactivating solution and diffusion equilibrium is established between the solution and model. On this occasion the ATP exerts its plasticizing (i.e., softening) action on the contractile structures and overcomes their rigidity, so that mechanical activity can thereafter take place. After transfer of the model into the same solution but at a higher temperature, intensive hydrolysis of ATP takes place and is accompanied by a motor response. This preliminary treatment of the model with cold reactivating solution is justified and is used with models which are relatively thick, such as those obtained from tissue culture films.

Only the basic principles of model making have been described here, but at the end of the book full details of the methods used to obtain models from various objects are given.

The principal experimental opportunities provided by models may next be examined. First, models can be used to identify the chemical substances at the expense of which a given contractile system works during contraction and relaxation, and to establish the environmental conditions which facilitate or, on the contrary, depress its work. They enable the mechanical parameters, the thermodynamics, and the kinetics of the motor response of an object to be studied under different environmental conditions and under the influence of various agents.

Models have revealed differences between the behavior of different objects and different contractile systems with respect to substances essential for performance of the motor act (Mg^{++}, Ca^{++}, ATP, and so on). The same substances induce a phase of contraction in some objects, a phase of relaxation in others, and a complete cycle of contraction and relaxation in yet a third group of objects.

It was on glycerinated models of myofibrils, by the study of their contraction under the action of ATP, that those observations were made by phase-contrast, polarization, and electron microscopy and by low-angle x-ray structural analysis which led to the creation of the "sliding theory" of the mechanism of muscle contraction (Hanson and Huxley, 1955; Huxley, 1968, 1969). The distribution of individual contractile proteins in the myofibril has also been studied on models (Hanson and Huxley, 1955; Perry, 1955; Aronson, 1965; Shtrankfel'd et al., 1966; Chaplain, 1970). By the use of models the mechanism of contraction was identified in the protozoan myoneme, which differs radically from the mechanism of contraction of the muscle fiber (Levine, 1956; Hoffmann-Berling, 1958; Seravin et al., 1965; Seravin, 1967). Work on models has elucidated the mechanisms of the successive motile responses taking place during mitosis. Without the use of models it would hardly have been possible to analyze these mechanisms and the differences between them. Even in such a continuous and, therefore, such an apparently single movement as the anaphase separation of the chromosomes, models have distinguished two interchanging processes with different mechanisms. The reason is that models enable the mechanical effect of the chemical reaction to be seen, and the morphology of the contractile apparatus to be examined by varying the chemical conditions under which it functions. This cannot be done in work with the living cell, with its selective permeability and its self-regulatory mechanisms, or in work with isolated contractile protein.

Results indicating the character and conditions of the work of models can reveal the similarities and differences between the energetic principles governing the motor activity of different cells and between the actual details of that activity. Comparison of different mechanisms of movement gives a better understanding of each of them and provides interesting information relative to the evolution of the motor function.

Models provide an experimental approach to the analytical study of regulation and coordination of the contractile function in the cell. This was the aim of experiments undertaken to study the action of "relaxation granules" isolated from muscles or from cells of connective-tissue cultures, or of soluble relaxation factors, on contractile models. Models can show whether a particular feature of the contractile function of a living cell is directly connected with the contractile system or whether it is due to other systems interacting with the contractile system. For example, if some particular feature of the motor act observed in the living cell is also found during the work of the model, this proves that it is an intrinsic feature of the contractile apparatus itself. If, on the other hand, the model does not possess this feature, it must be ascribed to systems present in the living cell but absent in the model.

The same approach can be used to study the site of action of substances stimulating the motor activity of the cell or, on the contrary, inhibiting it. Acetylcholine causes contraction of the living muscle. It has no action on the model fiber. Consequently, acetylcholine does not act on the contractile system, but does act on some other system, through which, i.e., indirectly, it activates the contractile system (Kory, 1950).

Comparison of the resistance of the contractile function of a living cell and of its model to a number of agents (inhibitors, changes of temperature, high hydrostatic pressure) has shown that in some cases a particular agent directly injures the contractile apparatus, while in others it depresses contractility in another way (Arronet, 1964a, 1964b, 1968; Arronet and Konstantinova, 1964, 1969).

The site of action of certain agents which increase the resistance of cell contractility to injury was determined in a similar way (Ganelina, 1962; Vorob'ev and Ganelina, 1963; Arronet, 1964a).

Models have also found applications in cytoecology. Phylogenetic differences in thermostability of the contractile cells of closely related species of poikilothermic organisms have been found to correspond to similar differences in the thermostability of the contractile apparatus of these cells. Like their living cells, models of the cells of a more thermophilic species possess greater thermostability (Aleksandrov and Arronet, 1956; Arronet, 1964b). This suggested the existence of phylogenetic differences in the thermostability of the contractile apparatus of cells — differences connected with the temperature ecology of the species and determined by it. Since the contractile apparatus of models is essentially a protein system, these results provided a weighty argument in support of the earlier conclusion that cell proteins of cold-blooded animals are adapted to the temperature conditions under which the species lives (Aleksandrov, 1952; Ushakov, 1956).

Contractile models have thus been applied in research to study not only the mechanism of contraction, but also other problems in cytophysiology.

It is to be expected that motile models of cells and muscles will yield new facts in the future in fields where they have already been successfully used and also in fresh fields which still await their application. There is no reason why this should not apply, in particular, to technology.

Chapter II

Properties of Various Motile Models

Photoplasmic movements are due to active shortening or active lengthening of structures. Both these events are connected somehow or other with ATP: they are either evoked or inhibited by it. Accordingly, cell movements can be classified as follows:

1. Contraction evoked by ATP;
2. Lengthening evoked by ATP;
3. Lengthening evoked by Ca^{++} ions and inhibited by ATP;
4. Contraction evoked by Ca^{++} ions and replaced by relaxation on addition of ATP. To these must be added:
5. Movements consisting of a single combination of contraction and lengthening;
6. Rhythmic movements of flagella, cilia, and certain muscles which make frequently repeated combinations of contraction and elongation.

In accordance with this classification, models of different types of protoplasmic movement will be described below.

1. CONTRACTION EVOKED BY ATP

Muscle Models

The model of the glycerinated muscle fiber was first obtained in Szent-Györgyi's laboratory, initially as longitudinal sections through glycerinated fibers (Varga, 1946), and later as carefully dissected bundles of glycerinated fibers (Szent-Györgyi, 1949). Szent-Györgyi worked with bundles about 1 mm thick. Weber (1951) showed, however, that accurate results can be obtained only by working with single fibers, for as the ATP diffuses into the bundle it is distributed along a diminishing gradient and, consequently, different fibers of the same bundle are under different conditions and contract with different force. In the center of a bundle 500 μ thick, corresponding approximately to the thickness of 10 fibers, the ATP concentration is zero. The reason for this is as follows. Fibers close to the surface hydrolyze ATP at a faster rate than it can diffuse from the medium into the bundle. As a result, ATP cannot reach the central fibers. Meanwhile the ATP concentration in the center of a single fiber 50 μ thick is only 1% lower than in the medium (Weber, 1951). Fibers not reached by ATP not only do not contract, but they prevent others from contracting for they are in a state of rigor very similar to rigor mortis.

What takes place in the muscle fiber during extraction converting it into a model? Phase-contrast investigations (Huxley and Hanson, 1954, 1960; Hanson and Huxley, 1955; Nayler and Merrilles, 1964; Abbott and Chaplain, 1966; Jagendorf-Elfvin, 1967; Sjöstrand and Jagendorf-Elfvin, 1967; Komissarchik and Shapiro, 1969, 1971; Ashhurst, 1969) have shown that the contractile elements of the model fiber have the same appearance as those in the living fiber (Figs. 1-4). The dimensions of the sarcomeres and their disks, the appearance and mutual arrangement of the thick and thin protofibrils, and the connections between them and also between the thin fibrils and Z-membranes remain visibly unchanged. The terminal cisterns of the sarcoplasmic reticulum and elements of the T-system also remain intact. In general, all the principal sarcoplasmic structures observable in specimens prepared from the living fiber except catecholamine-like granules can be seen in the model fiber. However, all membranes lose their sharp appearance and appear swollen and granular, with diffuse outlines. The mitochondria are swollen and sometimes ruptured, while elements of the sarcotubular system may also be broken. Nevertheless, even after keeping for 10 months in glycerol solution many mitochondria still remain unruptured. The nuclear membrane is much thinner, but the nucleus and nucleolus themselves are well preserved, with hardly any change in their electron density. Sometimes, admittedly, after glycerol extraction for several weeks, the sarcomeres are ruptured and the break is always formed in the I-disks.

The following observations have been made on the biochemical activity of glycerinated fibers.* Besides ATPase activity of actomyosin, and acetylcholin-estherase activity, fiber models also retain high activity of a number of enzymes connected with the mitochondria. Wilson and co-workers (1959), for example, found that glycerinated fibers utilize oxygen even after keeping for 4 months in glycerol solution. The surprising thing is that the intensity of oxygen absorption remained the same regardless of the presence or absence of a respiration substrate (succinate) in the medium. The presence of a functioning system of cytochromes in model fibers was confirmed by these workers spectroscopically; however, cytochrome c is partially extracted on glycerination.

Nayler and Merrilles (1964) also found a number of mitochondrial enzymes (succinate dehydrogenase, cytochrome oxidase) as well as the sarcoplasmic enzyme phosphorylase, in an active state in model fibers. The activity of these enzymes falls during keeping in 50% glycerol at -15°C, but it is still very high after 6 weeks and is very little different from the activity in fresh muscle, while after extraction for 10 months it is reduced by only 33-67%. Nayler and Merrilles consider that since the lysosomes are well preserved in models, the enzymes contained in them also probably remain in an active form.

These authors warn that glycerinated models of the muscle fiber must not be considered preparations consisting only of contractile proteins. They point out that in the model fiber there is a series of points on which tested substances can act although bypassing the contractile system. In other words, in their opinion preservation of the noncontractile systems detract considerably from the advantages of models of the fiber over the living fiber for the sake of which models are employed.

I do not think that the supporters of this view are justified in the degree of their skepticism. The work of the oxidative enzymes remaining intact in the model fiber must be regarded as isolated and unproductive work, incapable of affecting the contractile system. Strictly speaking it is not an oxidative system which remains, but its isolated, disconnected fragments. The discovery of a membrane potential in glycerinated fibers only 2-3 times smaller than that in the living fiber (Nayler and Merrilles, 1964)

---

*See Notes Added in Proof, page 167.

is more surprising. Nevertheless, it still remains a fact that the model fiber cannot respond by contraction to electrical stimulation. The reason probably is not that the model fiber can generate (although it must be added that this fact has never been confirmed*) only weaker potentials than the living fiber, but that the excitation—contraction coupling is broken in the model.

It seems to me pointless to continue this argument, for the following reasons. Successive extraction of muscles with glycerol and detergent solutions (Abbott and Chaplain, 1966) yields model fibers with a negligible number of mitochondria, virtually free from traces of noncontractile proteins, without oxidative phosphorylation and without respiration, but with complete integrity of their morphology and of the mechanical and biochemical properties of their actomyosin system; noncontractile ATPases, and those contractile membrane ATPases whose presence can sometimes distort the result of an investigation, are removed. This can be done on a very difficult object, namely the flying muscles of insects, in which the mitochondrial apparatus is particularly well developed. Leaving aside the arguments ventilated above, work must now be done to apply this method of extraction (see page 136) to other muscle and nonmuscle objects, and following Abbott and Chaplain by using cytochemical, biochemical, biomechanical, and electron-microscopic tests.

ATP has a twofold action on muscle models, whether they be actomyosin threads, models of myofibrils, or glycerinated fibers.

First, ATP plasticizes or softens the model, bringing it from a state resembling rigor mortis. Incidentally, the development of rigor mortis of a muscle is itself due to exhaustion of its ATP reserves. The plasticizing effect is the result of addition of ATP to the contractile material of the model, and not to hydrolysis of the ATP. Plasticization can also be produced by other polyphosphates, including inorganic, and by increasing the ionic strength of the medium by the addition of urea (Portzehl, 1952; Weber and Portzehl, 1952a, 1952b). It was probably plasticization of actomyosin threads under the action of ATP that Éngel'gardt and co-workers (1941) observed when working with threads to which a load was attached. ATP caused plasticization of these threads, they lost their rigidity, and the load stretch-

---
*See Notes Added in Proof, page 167.

ed them. The mechanism of plasticization consists of rupture of the cross-bridges formed by subfragments-1 of H-meromyosin between the myosin and actin protofibrils. Clearly the same phenomenon lies at the basis of the decrease in viscosity of actomyosin solution under the influence of ATP. Ca^{++} prevents plasticization.

Second, ATP provokes contraction of the fiber which it has plasticized. This effect arises only as a result of ATP hydrolysis, and the rate of hydrolysis must exceed 0.4 µg phosphorus/mg protein per minute. Contraction, by contrast with plasticization, can be caused only by ATP, and in the case of the fiber or myofibril model, by nucleoside triphosphates related to ATP, and also by ADP, which is first converted into ATP by adenylate kinase. Contraction by the action of ATP was first observed by Szent-Györgyi (1942-1942), who worked with freely floating actomyosin threads, i.e., with unloaded threads.

Usually the investigator has to consider both types of action of ATP on the model: plasticizing, followed immediately by contracting. However, by suppressing ATP hydrolysis, by Salyrgan,* for example, and thus blocking the contraction, the plasticizing action of ATP on the model can be isolated and in this way relaxation (isometric conditions) or lengthening (isotonic conditions) of the model can be obtained under the influence of ATP in a state of tension or contraction (Fig. 5). The model can also be plasticized by ATP against the background of EGTA, which removes Ca^{++} and thereby inhibits myosin ATPase/[(3-5) × 10^{-3} M ATP + 1 × 10^{-3} M EGTA]. In this way physiological relaxation of the muscle fiber can be simulated.

ATP in concentrations higher than 1 × 10^{-1} M plasticizes the fiber, but cannot be hydrolyzed and cannot therefore cause contraction of the model; on the contrary, the model relaxes or lengthens (Fig. 6). These ATP concentrations are called superoptimal. The effect of the superoptimal ATP concentration is considered (Weber, 1958; Hoffmann-Berling, 1959) to be a case of enzyme-substrate inhibition. If this is so, there is no difference in principle between plasticization by ATP in normal concentration against the background of Salyrgan and plasticization caused by

---

*Salyrgan: salicyl-(3-hydroxymercuri-2-methoxypropyl)-amido-orthoacetate; proprietary name for mersalyl, with the action of a sulfhydryl poison.

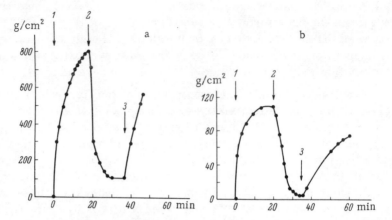

Fig. 5. Development of tension in a glycerinated model of a psoas muscle fiber and an actomyosin thread from rabbit muscles by the action of ATP and its relaxation by the action of Salyrgan (after Portzehl, 1952). a) Model of fiber (1.7×10^{-3} M ATP), b) actomyosin thread (2.5×10^{-3} M ATP). Abscissa, time of experiment; ordinate, tension developed by preparation. Arrows: 1) addition of ATP, 2) addition of 1×10^{-4} M salyrgan, 3) termination of action of salyrgan by high dilution and addition of 2×10^{-2} M cysteine.

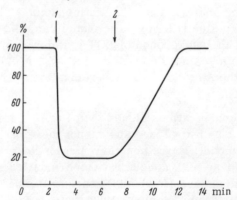

Fig. 6. Contracting (1) and lengthening (2) action of ATP on muscle fiber model (from Weber and Portzehl, 1952a; after Bozler, 1951). Abscissa, time of experiment; ordinate, change in length of fiber model (in % of initial length). Arrows: 1) addition of 3.5×10^{-3} M ATP to medium, 2) addition of 1×10^{-1} M ATP.

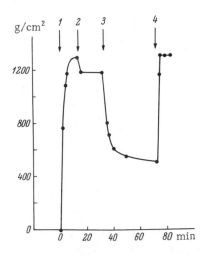

Fig. 7. Development of tension of glycerinated model of rabbit psoas muscle fiber by the action of ATP and its relaxation by the action of inorganic pyrophosphate (from Portzehl, 1952). Abscissa and ordinate as in Fig. 5. Arrows: 1) addition of 2×10^{-3} M ATP, 2) rinsing, 3) addition of 2.5×10^{-3} M pyrophosphate, 4) addition of 3×10^{-3} ATP.

ATP in superoptimal concentration: in both cases the essential feature is the suppression of ATP hydrolysis.

Since inorganic polyphosphates can also plasticize muscle models, they can be used to abolish the state of contraction or tension of the model developing by the action of ATP in optimal concentration, as a result of which the model lengthens or relaxes (Fig. 7). The following conditions are essential for contraction of all types of muscle models (Weber and Portzehl, 1952a; Weber, 1958).

1. The presence of ATP in the medium, in a concentration not less than 1×10^{-3} M; the optimal effect is given by ATP in a concentration of 4×10^{-2} M. In the case of myofibrils and fiber models, ATP can be replaced by GTP, UTP, CTP, ITP, and ADP (Fig. 8).

2. The presence of magnesium and calcium ions in the medium in low concentrations: over $1 \times 10^{-6} M$ Ca^{++}, optimally $1 \times 10^{-4} - 1 \times 10^{-2}$ $M Mg^{++}$. Models of different fibers differ somewhat in their requirement of Mg^{++}. For instance, contraction of models of skeletal striped muscle fibers of different animals and also smooth-muscle fibers of insects requires magnesium ions in concentrations below 1×10^{-4} M, contraction of heart muscle fiber models requires $(1-2) \times 10^{-3}$ $M Mg^{++}$, and models of vertebrate smooth-muscle fibers require $(3-4) \times 10^{-3}$ $M Mg^{++}$ (Filo et al., 1965;

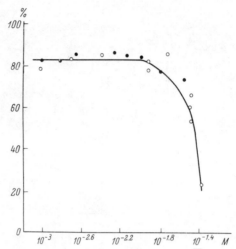

Fig. 8. Degree of shortening of myofibril models by the action of ATP (circles) and ITP (dots) (from Portzehl, 1954). $\mu = 0.12$, pH 6.9. Abscissa, ATP or ITP concentration; ordinate, change in length of models (in % of initial).

Rüegg, 1968). Magnesium ions not only activate ATP hydrolysis, but also stimulate ATP adsorption by actomyosin structures, which gives a plasticizing effect. Contraction of the models, although not maximal, also takes place in media to which no Mg^{++} is added. Ca^{++} formerly was never added to media intended for the reactivation of muscle models. The reason was that the small quantities of Mg^{++} and Ca^{++} which are always present in the model itself and in medium to which they have not been specially added are sufficient for the work of the muscle model.

In nonquantitative investigations, Ca^{++} need not be added to media reactivating muscle models: in commercial ATP (samples obtained from the firms "Sigma" and "Pabst") were investigated, to 500 moles ATP there is 1 g-ion Ca^{++} (Seidel and Gergely, 1963); this means, for example, that in 5×10^{-3} M ATP, there is 1×10^{-5} g-ion Ca^{++}/liter. If the effect of Ca^{++} on models is to be studied, the solutions are first freed from it by means of EGTA or a resin of the Chelex-100 or Dowex-50 type, and Ca-EGTA buffers are used. In this way it has been shown on models of striped and smooth muscles that Ca^{++} is absolutely essential for the work of the models, and that the tension of the model fiber and also its

ATPase activity are directly dependent on the Ca^{++} concentration (Schädler, 1967). Ca^{++} in concentrations of $1 \times 10^{-7} - 1 \times 10^{-6}$ g-ion/liter gives 50% activation of muscle models from a wide range of different objects. Schädler's work is also interesting because he recorded the tension and ATPase activity of the models simultaneously in the same experiment.

That Ca^{++} and Mg^{++} are essential for contraction is clearly demonstrated if all traces of them are specially removed from the medium (Ca^{++} by means of EGTA, Mg^{++} by means of EDTA): this leads to disappearance of the contractile response of the muscle model to ATP. In fact, the $(Mg-ATP)^{--}$ ions are the ones that act to evoke mechanical response.

3. The rate of hydrolysis of the nucleoside triphosphate must be not less than 0.4 μg phosphorus/g protein per minute. This requires a sufficiently high concentration of ATP throughout the thickness of the model (the diameter of the preparation must not exceed 75 μ for striped muscle models and 100 μ for smooth muscle models, because of the lower ATPase activity of the latter). The actomyosin system of the model must be in the native state. The muscle fibers must be extracted long enough to destroy the sarcoplasmic reticulum sufficiently to ensure that its membranes can no longer act as a relaxation system; the ATPase incorporated in the membranes of the sarcolemma and sarcoplasmic reticulum must be inactivated since otherwise it will hydrolyze the ATP reducing its diffusion to the contractile elements.

Probably glycerinated and later detergent-treated models are more suitable in this respect also. It is possible that they can also be thicker.

4. The ionic strength and pH of the medium must be within physiological limits: ionic strength about 0.15, pH between 6.8 and 7.8 (optimum 7.4-7.6).

It will be understood that, of all conditions listed above, those connected with the need for disruption and inactivation of the relaxation system apply only to models of the fiber, while those connected with the thickness of the model do not apply to model myofibrils.

Muscle models do not contract under the influence of ATP if "relaxing factor" in present. This factor was first mentioned by

Marsh (1951, 1952) and Bendall (1952, 1954). They found that muscle homogenate contains an extractable agent capable of suppressing the contractile response of glycerinated muscle preparations to ATP. This factor was known for many years as "Marsh's factor" or "Marsh—Bendall factor." Portzehl (1957) later showed that this factor, in granule form, can be sedimented from homogenates by centrifugation. "Relaxing granules" or "vesicles" were shown to be a product of mechanical disintegration of elements of the sarcoplasmic reticulum, which permeates muscle fibers. These granules themselves are not the relaxing agent, but react with ATP and Mg^{++} to form it. As a result of this reaction, which takes place in the membrane wall of the granules and is accompanied by rupture of the high-energy bond of ATP, calcium ions are actively transported inside the granules (vesicles). As a result there is a sharp decrease in the calcium ion concentration outside the vesicles, near the actin and myosin protofibrils, and the mechanochemical reaction between the contractile elements of the muscle and ATP does not take place. The role of the relaxing granules thus consists essentially of removal of calcium ions from the contractile elements and their accumulation inside the vesicles.*

It is very important to note that the relaxing effect of granules on the muscle model can be completely abolished by the addition of Ca^{++} in a concentration greater than 1×10^{-6} g-ion/liter.

---

*There is evidence that relaxing granules, reacting with ATP and Mg^{++}, synthesize a soluble relaxation cofactor. This binds Ca^{++} actively, thereby relaxing the fiber, i.e., the result of its action is the same as that of the insoluble factor — the relaxing granules.

A number of papers dealing with the problem of muscle relaxation and, in particular, relaxation of heart muscle, are included in the Proceedings of the 48th Annual Conference of the Federation of American Societies of Experimental Biology (Chicago) (see: Federation Proceedings, Vol. 23, No. 5 (1), pp. 885-939 (1964)). These papers examine the macrostructure and microstructure of the sarcoplasmic reticulum and T-system in the muscle fiber; the role of Ca^{++} in the regulation of the contraction—relaxation cycle of skeletal and cardiac muscle; the problem of the existence of a soluble relaxing factor; properties of relaxing granules, the connection between relaxation and the calcium pump; the role of phospholipids in ATPase activity of the membranes of the sarcoplasmic reticulum and its ability to trasport Ca^{++}. Many of the investigations described in these papers were carried out with the aid of models.

Under these conditions, despite the presence of relaxing granules, the muscle model begins to contract under the influence of ATP.

The facts described above, obtained by work using muscle models and a preparation of relaxing granules, simulate processes which evidently must take place during physiological contraction and relaxation of muscle. The view has been expressed (see, for example: Weber, 1960; Hasselbach, 1963, 1964; Krolenko, 1965; Ivanov, 1968) that the sequence of events during the work of the normal muscle fiber conforms to the following scheme. In a resting state calcium ions accumulate in the elements of the sarcoplasmic reticulum. Their departure from it is blocked by the polarized state of the reticulum membranes. The Ca^{++} concentration in the immediate vicinity of the contractile protofibrils is below 1×10^{-6} g-ion/liter. In such a low Ca^{++} concentration hydrolysis of ATP does not take place and there is no contraction despite contact between ATP and the contractile proteins; the ATP has only a plasticizing action. Excitation reaching the end-plates along the nerve moves over the sarcolemma of the fiber depolarizing the membranes, spreads along the tubules of the T-system, and depolarizes the membranes of those elements of the sarcoplasmic reticulum which are near to the T-system. As a result, calcium ions are able to pass through the depolarized reticulum membranes from its large cisterns and longitudinal canals, where they are in a concentration of the order of 1×10^{-4} g-ion per liter, and they rush headlong in the direction of the concentration drop in the immediate neighborhood of the protofibrils. As soon as their concentration near the contractile elements reaches 1×10^{-6} g-ion/liter, the "ignition" mechanism is triggered, and this sets in motion the mechanochemical reaction involving the participation of the myosin and actin protofibrils and ATP, the latter undergoing hydrolysis in the process. Contraction of the myofibril, fiber, or muscle takes place (Ebashi et al., 1969).

The cause of relaxation of a muscle and abolition of its contracted state is considered to be the return movement of Ca^{++} from the sarcoplasm through the membrane of the reticulum into its central cisterns, canals, and vesicles. This movement of Ca^{++} takes place by active transport at the expense of ATP and with the necessary participation of Mg^{++} as activator. When the Ca^{++} concentration next to the protofibril falls below 1×10^{-6} g-ion/liter, myosin-dependent ATP hydrolysis stops through lack of activator, the contracted state of the myofibrils is abolished, and since the plasticizing action of the ATP continues, elastic forces bring about relaxation and lengthening of the myofibrils.

Autoradiographic investigation of the rabbit muscle fiber using Ca^{45} has revealed the actual areas of the sarcoplasmic reticulum from which calcium ions are discharged to the myofibrils in response to excitation. These areas were shown to be close to tubules of the T-system. The return flow of Ca^{++} into the canals of the reticulum takes place in areas remote from the Z-membranes and, correspondingly, from the tubules of the T-system. Tice and Engel (1966) found that Mg-dependent ATPase is localized in these same areas of the sarcoplasmic reticulum.

Ridgway and Ashley (Ridgway and Ashley, 1967; Ashley and Ridgway, 1968) have recently published the results of their elegant experiments in which they were able to see Ca^{++} leaving the cavities of the sarcoplasmic reticulum for the protofibrils by direct vision. These workers injected jellyfish protein (aequorin), which emits visible luminescence when the Ca^{++} concentration in the medium exceeds 1×10^{-6} g-ion/liter, into a giant muscle fiber of Balanus. This protein impregnated the myofibrils without entering the canals or cisterns of the reticulum; it did not luminesce in the resting fiber. Electrical stimulation of the fiber caused the development of a depolarization potential, which was followed by the appearance of luminescence, and after a short latent period the fiber contracted. In these experiments calcium ions, in the same concentration (1×10^{-6} g-ion/liter) induced outside the reticulum by the electrical stimulation, concurrently activated luminescence of the indicator protein and the mechanochemical reaction of actomyosin with ATP. As a result of these ingenious experiments it was literally possible to see Ca^{++} moving from the canals of the reticulum toward the protofibrils. If weak stimulation was used a subthreshold potential was generated, at which neither luminescence nor contraction took place: the stimulation was insufficient to produce the depolarization of the reticulum membranes which would have allowed the calcium ions to escape through the pores in these membranes and to activate luminescence of the aequorin and contraction of the myofibrils.

It is very important to note that other, noncontractile, functions of the cell which take place at the expense of the energy of

ATP, such as the discharge of secretion from gland cells, or the biological luminescence of various objects, are also activated by Ca^{++}, and that in many cases the process is triggered by the appearance of Ca^{++} in that same concentration of 1×10^{-6} g-ion/liter near the appropriate structure. This is a fact of definite evolutionary interest (Krolenko, 1965).

Experiments with models and preparations of relaxing granules, and also with preparations of soluble relaxing factors can thus help to analyze the mechanism of relaxation of muscle (and, as we shall see later, of other cells) by reproducing relaxation in a simplified contractile system and studying the effect of various factors and conditions on this process. Since mechanical work of any kind is a combination of contraction and relaxation, elucidation of the mechanism of relaxation is just as important as elucidation of the mechanism of contraction.

According to another point of view, granules cause relaxation by liberating a certain cofactor which penetrates into the fiber and prevents it from contracting. This assumption provides an explanation for the following facts obtained with models of structures other than muscles, yet of general importance. A preparation of relaxation granules (which can be obtained either from muscles or from connective-tissue cells), added externally to a cell model, does not act uniformly on the body of the model in contact with the granules, but only at places where the relaxation system of the actual cell from which the model was made is located (Kinoshita et al., 1964). It is claimed (Hoffmann-Berling, 1964) that the model's own relaxation system, which was disrupted by extraction, is thereby reactivated by a cofactor liberated from the relaxation granules (see page 30 and the footnote to it). Facts of this nature will be examined more fully in the section on models of cytokinesis (page 81).

A fiber model developing tension under isometric conditions under the influence of ATP + Mg^{++} is relaxed by the simple removal of these substances from the medium. However, the same model, if it contracts under isotonic conditions under the influence of the same ATP + Mg^{++}, does not lengthen after their removal (Watanabe et al., 1960). Watanabe and co-workers describe several experimental facts showing relaxation of the fiber model under the influence of various chemical agents and differences between the reactions of models to these agents under isometric and isotonic conditions. Lengthening of relaxation of the model fiber, when contracted or in tension as the result of treatment with ATP + Mg^{++} takes place if EDTA is added to the medium in a concentration of the order of 1.5×10^{-3} M. The relaxation can subsequently be abolished and tension or contraction of the fiber again produced by means of Ca^{++}, Sr^{++}, Cd^{++}, Mn^{++}, Zn^{++}, Co^{++}, or Ni^{++}, but not by Mg^{++}. It is significant that Ca^{++} and Sr^{++}, which cause contraction (it is better to say abolish the relaxation produced by EDTA) in concentrations much below those capable of binding the whole of the EDTA present; Mn^{++}, Co^{++}, Ni^{++}, Cd^{++}, and Zn^{++} can abolish this relaxation only if their concentration exceeds that of the EDTA. In experiments under isometric conditions the tension produced by Cd^{++} or Zn^{++} is spontaneously replaced by fresh relaxation which, in turn, can be prevented by the addition of cysteine to the medium. In experiments under isotonic conditions, on the other hand, shortening of the model by Cd^{++} or Zn^{++} is not replaced by spontaneous lengthening.

Relaxation of the glycerinated fiber when preliminarily contracted by the action of ATP + Mg^{++} can also be produced by inorganic pyrophosphate. Under these conditions, as Watanabe et al. found in their experiments mentioned above, a fiber under isometric conditions relaxes quickly and the concentration of pyrophosphate required is low, of the order of 3×10^{-3} M; under isotonic conditions, on the other hand, relaxation of the fiber takes place slowly and only by the action of pyrophosphate in high concentrations, namely $(2-3) \times 10^{-2}$ M.

So far we have dealt mainly with the glycerinated muscle fiber, although this is not the only model of muscular contraction. What are the distinguishing features of other muscle models? How do they resemble and differ from those we have examined? How truthfully do the mechanical properties of the various models reflect the contractile properties of the living muscle fiber? What are the advantages and disadvantages of some models over others? How should the experimenter be guided when choosing models for his experiments?

Isolated glycerinated and nonglycerinated myofibrils (Schick and Hass, 1949; Perry, 1951; Hanson, 1952; Hasselbach, 1952; Perry and Horne, 1952; Portzehl, 1954; Ulbrecht and Ulbrecht, 1957; Jewell et al., 1964; Reedy et al., 1965; Jewell and Rüegg, 1966; Solaro et al., 1971) provide a "purer" model than glycerinated fibers for the study of the working cycle of muscle (Figs. 3 and 4). Glycerinated myofibrils have neither sarcolemma nor remains of the sarcoplasma nor sarcosomes, so that the doubts expressed by some authorities regarding the autonomy of the contractile system in the model fiber (see page 23) cannot be applied to model myofibrils.

Besides contractile proteins, nothing is left in these models but 5-adenylate deaminase (which, incidentally, can be removed) and small amounts of adenylate kinase (this can be washed out by salt solution with ATP, but under these circumstances the myofibrils will naturally contract). Because they contain traces of adenylate kinase, isolated myofibrils will contract in the presence of ADP as well as of ATP. Their disadvantage is that the force of contraction which they develop cannot be measured. They are also difficult to use for studying relaxation. However, the degree of shortening and its velocity under various conditions (measured from motion pictures) have been studied successfully (Portzehl, 1954). Models of myofibrils are recommended for work requiring investigation of the staining reaction of normally oriented contractile proteins (see page 116) to various agents, for unlike fiber models, the glycerinated myofibril consists practically entirely of these proteins, and just as in the fiber model, their protofibrillary organization is undisturbed. The best objects from this point of view are the myofibrils from glycerinated and detergent-treated fibers (Abbott and Chaplain, 1966).*

Isolated protofibrils and resynthesized myosin and actin filaments have also been used.

Isolated protofibrils (thick and thin threads of myofibrils, or myofilaments) are obtained from glycerinated myofibrils by homogenization in a medium producing plasticization, i.e., rupture of cross-bridges of subfragments-I of H-meromyosin between the myosin and actin protofibrils). Either pure preparations of myosin and of actin protofibrils or a mixture of the two can be obtained (Huxley, 1963). This model marks the next stage in the experimental "strip-tease" of the contractile apparatus of the muscle. The myosin protofibrils have a diameter of 100-200 Å, a length of 1.5-1.6 μ, and conical ends with projections serving as bridges joining them to the subfragments-1 of H-meromyosin, which are not found in the middle part of the protofibril. Actin protofibrils are 60-70 Å in

---
*See Notes Added in Proof, page 168.

diameter and 0.5-1.0 µ long (longer protofibrils have been found as a result of polymerization of the normal protofibrils); they are frequently combined into double bundles with a central connection — these are essentially isolated I-disks with a Z-membrane. Protofibrils of the same length are obtained from resting and contracted muscles; the length of both types of protofibrils is unchanged by the action of $ATP + Ca^{++} + Mg^{++}$. These are important facts in support of the sliding filament theory.

Synthetic myosin and actin filaments, closely resembling natural protofibrils in their structure, have also been obtained from protein solutions. Myosin filaments are formed by reducing the ionic strength of a solution of myosin monomer (0.5 M KCl) to 0.15-0.17, at pH of strictly 7.2-7.5 (Huxley, 1963; Josephs and Harrington, 1966; Kaminer and Bell, 1966). Depending on the rate of dilution of the protein (dialysis for 24 h or dilution with water in 1-3 min) the filaments vary in size, reaching 150 Å × 2-12 µ, but more usually 60-100 Å × 0.5-1.0 µ, which is close to the size of the thick filaments of the myofibril. Myosin filaments are bare in their central part (0.15-0.2 µ), while their conical ends carry projections, or heads, of H-meromyosin, which are potential crossbridges. The myosin molecules (20-40 Å × 0.15 µ) in them are polar in their polymerization: they lie with their heads toward the ends and their tails toward the middle. These filaments are very similar to the thick protofibrils of the normal sarcomere or smooth-muscle fiber.

Actin filaments (F-actin) are obtained from an aqueous solution of G-actin by adding ATP increasing the ionic strength to 0.1. Polymerization takes place with stoichiometric addition of ADP and removal of P_{inorg} (Szent-Györgyi, 1960).

Natural and artificial protofibrils of myosin and actin can be used to investigate the supramolecular organization of the elements of the myofibril and the morphology of the elementary motor act, with the expectation that it will in the future be possible also to study its biochemistry and energetics. Very probably, by means of models such as these, the decisive facts which will give the key to the mechanism of the elementary motor act, in what can be termed the last approximation, will be obtained. Naturally, preparations of subfragments-1, troponin, and tropomyosin will also be used in conjunction with them.

An essential difference between F-actin filaments and natural thin protofibrils must be mentioned, although it is not always emphasized. Natural actin protofibrils have a strictly definite length (about 1 μ for cross-striated muscles, counting the length of the filament as the distance along one side of the Z-membrane, i.e., half of an I-disk) and molecules of other proteins — tropomyosin and troponin — are contained inside two twisted actin helices. If actin filaments are prepared artificially, a mixture of filaments of different lengths, some longer and some shorter than the natural, is obtained. Admittedly, by choosing the ionic strength and pH of the medium, filaments of approximately the natural size can be obtained, but the result will still be a polydisperse mixture although the degree of polymerization of G-actin is within a not very wide range (Kaminer and Bell, 1966). The possibility is not ruled out that it is tropomyosin which regulates the length of the natural actin protofibril, while, at the same time, it codes and defines the degree of polymerization of G-actin. In artificial filaments, on the other hand, there is no tropomyosin. There is likewise no troponin, and this is an important protein, the allosteric regulatory subunit of the natural actin protofibril; in the absence of Ca^{++} troponin blocks interaction between the actin and myosin protofibrils, and in the presence of Ca^{++} this prohibitive action of troponin is abolished (Ebashi et al., 1969; Huxley, 1969). In the future it will perhaps be possible to obtain artificial actin filaments which include tropomyosin and troponin in the natural amounts and organized in the natural manner. In addition, it would be useful to obtain actin-tropomyosin filaments and, separately, tropomyosin-troponin preparations, and to be able to cause them to function. Such models would help to explain the way in which calcium regulates the ATPase activity of H-meromyosin and triggers the mechanochemical reaction. (See also the section on preparation of isolated protofibrils in Chapter IV.)

The next stage along the path of simplification of the muscle contractile apparatus is the actomyosin thread. In threads obtained by Szent-Györgyi's (1942) method the actin and myosin molecules are not regularly oriented and the protein concentration is low.

The structure of actomyosin threads prepared by Szent-Györgyi's method has been investigated under the electron microscope (Beck et al., 1969). They contain a network of irregularly aggregated filaments. On contraction of the thread by the action of ATP the network becomes closer. Treatment of actomyosin threads with glycerol has no effect on their fine structure. A similar picture is found in glycerol-treated vertebrate smooth-muscle fibers, in which only the fibrils at the periphery of the model fiber are regularly oriented and parallel to each other; elsewhere a network of fibrils can be seen and it also becomes denser under the influence of ATP. The myofilaments in glycerol-treated invertebrate smooth-muscle fibers are more regularly oriented. Myosin protofibrils of vertebrates are similar to paramyosin protofibrils of invertebrates (in both cases, smooth muscles are implied). Unfortunately, no electron-microscopic study has been made of filaments presumed to have a more regular structure, namely those prepared by Portzehl's method, i.e., by drying under tension, and by Hayashi's method, i.e., prepared from films. However, such an investigation has been carried out on stretched strands of myxomycete plasmodium (Kamiya, 1968). To obtain longitudinal orientation of the internal structures of the plasmodium, strands cut from it were hung vertically for 1 h in a humid chamber, and a load of about 20 g/cm^2 attached to its bottom end, stretching the strip: this method is similar to that used by Portzehl for actomyosin threads. Electron-microscopic study showed that the fibrils in glycerinated models prepared from these living stretched strands are oriented parallel to the axis of the strand, and not arranged haphazardly as in models made from unstretched strands. Models from stretched strands contract in a longitudinal direction and can even oscillate rhythmically, repeating a contraction—relaxation cycle over and over again in the same way as stretched but unglycerinated strands. Unstretched models, on the other hand, contract only laterally under the influence of ATP.

Portzehl (1951) suggested a method of increasing the protein concentration in a filament uniformly throughout its volume (by drying threads soaked in glycerol slowly, in a cold, humid chamber), and of orienting its component protein particles more regularly (by stretching the threads). Because of the increased concentration of actin and myosin, the strength of interaction between their molecules in such a thread is high. This ensures relatively high breaking strength and a high degree of tension (contraction) developing under the influence of ATP, so that tensometric measurements can be carried out on such filaments. Hayashi (1951, 1952) introduced filaments made from films, made by compressing actomyosin monolayers from the side and rolling them into a filament, into scientific practice. Protein molecules in the monolayer are arranged mostly uniformly, and this orientation is maintained in filaments made from the film: Hayashi's photomicrographs show the fibrillary structure of the monolayer filament (Fig. 9). According to Hayashi's measurements, these filaments do not lose water during contraction. Their contraction

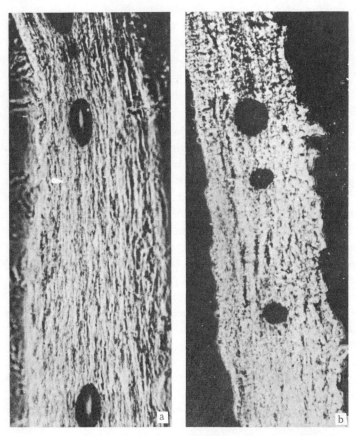

Fig. 9. Monolayer filaments from rabbit actomyosin (from Hayashi, 1952). (a) Uncontracted filament; (b) filament contracted under the influence of ATP.

is anisodimensional, i.e., it takes place only in the direction of the long axis.

Evidence and arguments have been adduced by Ivanov to show that stretched filaments (filaments prepared by Portzehl's method), like those made from monolayers, contain a high proportion of denaturated components. He also considers that they contract by a mechanism of entropy: "on account of the elastic forces of stretched structures of partially denatured protein" (Ivanov and Yur'ev, 1961, p. 101). According to Ivanov, the role of ATP is essentially to induce syneresis of undenatured parts of the filaments, setting free the elastic forces of the denatured segments. In that case contraction of stretched or monolayer filaments, on the one hand, and contraction of the model fiber or of Szent-György's freely floating threads, on the other hand, are based on different processes which are similar only in their external effect. However, the coincidence between the thermodynamic and kinetic characteristics of contraction of both groups of models suggests a common mechanisms of contraction in both cases.

Actomyosin gel, in which the actin and myosin molecules are not in regular order or orientation, is a more simplified model still. Under conditions common to contraction of all muscle models, it superprecipitates: it expels water and forms dense aggregates (Szent-Györgyi, 1941, 1942). It must be noted that water is also expelled by other models including the model fiber during contraction. Admittedly, when the latter contracts, its diameter increases, but not sufficiently to maintain its precontraction volume. Ivanov (Ivanov and Yur'ev, 1961) considers that monolayer filaments also lose water on contraction. However, all models differ in this respect from the living fiber, in the myofibrils of which, even if water is expelled during contraction, it is not expelled outside the structures themselves. Isolated myofibrils, when contracting under the influence of ATP and thus changing into globular form, are also dehydrated. As a result, their sedimentation constant and also the volume of the residue on centrifugation are sharply altered. This fact is made use of when the degree of contraction of model myofibrils under different experimental conditions is estimated (Bendall, 1969). The reason for this difference is evidently not that the mechanism of contraction is different in the muscle and its models, but that in the latter the properties of the membranes are disturbed.

Finally there is the simplest model, a streaming solution of actomyosin, in which the molecules are to some extent oriented. Important mechanochemical studies have also been undertaken on this model (Needham et al., 1941; Needham, 1950; Vorob'ev and Kukhareva, 1965).

For a review of the results obtained with solutions of actomyosin, actin, myosin, its components — and also of tropomyosin and troponin, see the surveys by Poglazov (1966), Perry (1967), Ebashi et al. (1969), and Bendall (1969).

I must now mention another two muscle models to be added to the range of models available for studying muscle function. These are models for studying systems and characteristics of excitation of muscle and the transmission of excitation to the contractile system.

First there is the muscle fiber from which the sarcolemma has been removed, a preparation studied by the Japanese physiologist Natori and described in detail by Costantin and Podolsky (1967). These fibers are unable to generate and propagate excitation over their surface. Under experimental conditions activation of the myofibrillary system of the bare fiber, however produced, takes place by depolarization of the sarcotubular system and by a decrease in the membrane potential of this system relative to the potential of the perifibrillary medium. This preparation can be used to study the electrical properties and permeability of the sarcotubular system and the transmission of the stimulus to the myofibrils.

Second, there are muscle fibers in which the sarcotubular system has been anatomically excluded or reversibly disrupted (Eisenberg and Gage, 1967; Gage and Eisenberg, 1967). In a fiber which has been kept for some time in Ringer's solution containing 3.5-4% glycerol, and then transferred into pure Ringer's solution, as the result of this transfer the tubules of the T-system swell and the connection between them and the sarcolemma is broken. This separation of sarcolemma and T-system is reversible. The preparation preserves all the properties associated with normal functioning of the sarcolemma membrane. In such a fiber a normal action potential can be evoked, but it is not followed by contraction because transmission of excitation to the membrane of the T-system and beyond cannot take place. In addition, the depolarization which normally arises immediately after the single action potential, and the slow changes in potential usually produced by prolonged hyperpolarization of the fiber do not arise: both these phenomena are associated with the T-system of the fiber.*

The investigator thus can avail himself of models for the isolated study of permeability, of excitation of the muscle fiber and its transmission to the myofibrillary system, and, finally, the actual working act producing the movement. Isolated mitochondria (sarcosomes) and sarcoplasm enzymes, which can be used to study individual components and the whole system of metabolic reactions of the muscle, complete the list of objects available for the analytical study of muscle functions.

Let us now turn to the contractile models themselves. Conditions leading to contraction of actomyosin threads, myofibrils, and fiber models are identical. Threads made of actomyosin from skeletal muscles, like model fibers, must not exceed 75 μ in diameter. The optimal pH, ionic strength, ATP concentration, and Mg^{++} requirement are the same for all types of muscle models and they agree, moreover, with the conditions in the living muscle fiber. The threads are more specific than the model fiber in their substrate relationships: the fiber, which still retains its adenylate kinase, will also contract in response to ADP, but the thread reacts by contraction only to ATP. As Table 1 shows, model fibers develop the same tension (contraction) as the corresponding muscles. The tension of the threads during isometric contraction is substantially lower, presumably because of the lower protein concentration in the thread. Moreover, the thread cannot develop a greater tension, for otherwise it would break.

The living skeletal muscle shortens during isotonic contraction at the most by 40-50% of its initial length. Smooth muscle shortens more than this — by 80% of its resting length. The degree of shortening of fiber models is equally considerable, whether obtained from smooth muscle (Ulbrecht and Ulbrecht, 1952) or from skeletal muscle (Annemarie Weber, 1951): they can shorten by 85%, i.e., to 15% of their initial length. Threads can also shorten to a greater degree than living muscle, but not quite so much as fiber models — by 70%. The degree of shortening of all types of

---

*See Notes Added in Proof, page 168.

TABLE 1. Maximal Tension Which Muscles and
Their Models Can Develop during Isometric
Contraction (kg-wt/cm^2)

Object	Living muscles	Fiber muscles	Actomyosin threads dried and stretched as described by Portzehl (1951)
Skeletal muscles:			
rabbit	5	4	0.2
frog	2	3	—
Smooth muscles:			
white adductor of Anodonta	4.5	2	0.2
layer of longitudinal fibers of the rectus abdominis muscle of a cow	0.7	0.6	—

Note: Table compiled from sources in the literature.

muscle models is thus greater than that of living skeletal muscles; for all models obtained from either smooth or cross-striated muscles it is close to the degree of maximum shortening of living smooth muscles.

The working efficiency of the model, or the ratio between the energy liberated by hydrolysis of ATP and the work done — is somewhat lower for models than for muscle, but it is of the same order.

Fiber models and threads contract at a slower rate than living skeletal muscles. This is understandable: time is required for ATP to penetrate along the radius of the fiber model or thread to its central axis, and the outer layers of the model begin to contract at the beginning of this period while the central layers, which are still rigid, resist the contraction. For this reason, it is customary to measure the rate of shortening of a thread or fiber model in the following manner. The muscle is placed in an isometric state in reactivating solution and left in it for a short time during which the model becomes permeated with ATP and tension develops. The model is then quickly released and the rate of its shortening measured at that moment. However, even if this "more refined" recording method is used the rate of contraction of model fibers and threads is slower than that of living muscle. The reason is not only the thickness of the models, impeding diffusion of ATP, but also the reduced ability of their enzymes to hydrolyze ATP: in skeletal muscle models the rate of ATP hydrolysis is only 0.8 μ mole phosphorus/sec per gram protein, compared with 4 μmoles for living muscle. The rate of contraction of a thin model, such as glycerinated myofibrils, under the influence of ATP is the same as the rate of contraction of the living muscle (Portzehl,

It is interesting to note that all these contraction indices for models, which differ from the corresponding indices for living cross-striated muscles, agree with those for living smooth muscles. These indices are the higher degree of shortening of the models and the slower rate at which it takes place. In the living skeletal muscle a higher degree of shortening (up to 80% of the initial length) and a slower rate of contraction or development of tension can also be obtained. Both are characteristic of the living skeletal muscle in the overexcited or delta-state (Ramsey and Street, 1940). Living skeletal muscle behaves under these circumstances just like smooth muscle or the model of any muscle. Weber and Portzehl (1952a) consider that smooth muscles have preserved their phylogenetically primitive properties: a high degree and low rate of shortening. During evolution skeletal muscle has acquired the ability to contract strongly and rapidly, at the expense of ability to shorten by a high degree. In models of skeletal muscles both the rate of shortening and its maximum degree are at the phylogenetically old level which is still characteristic of smooth muscles. Consequently, the increased rate of contraction and the lower degree of maximal contraction characteristic of cross-striated muscles are connected with certain systems no longer present in the models.

Other characteristics of contraction common to the fiber model, the actomyosin thread, and the living muscle can be listed. This applies, for example, to Hill's "quick-release" phenomenon, which is as follows. If a muscle in a state of isometric contraction is suddenly freed and allowed to shorten by a certain amount, at the moment of shortening, before the muscle has acquired a new, fixed length, its tension disappears completely, and not until it is again in the isometric state (at the new length) does the tension develop again. This phenomenon has been reproduced in fiber models and actomyosin threads from both cross-striated (Annemarie Weber, 1951; Portzehl, 1951) and from smooth (Ulbrecht and Ulbrecht, 1952; Dörr and Portzehl, 1954) muscles.

In living muscle, in threads, and also in fiber models, birefringence decreases equally during contraction (Ströbel, 1952).

The work of a model of the rabbit psoas muscle fiber, like the work of the muscle itself, obeys Hill's formulas expressing rate of contraction as a function of load (Ulbrecht et al., 1954).

This comparison of the character of contraction of the living muscle and its model thus shows that the course of contraction of the models is in fact the same as the course of contraction of the living muscle. It can accordingly be concluded that muscle contraction can justifiably be investigated with the aid of models.

The most important results obtained with fiber models are those related to the question of whether chemical agents causing contraction, and also agents which influence the character of contraction, in fact act directly on the contractile system of the muscle. Many substances stimulating muscle contraction have been found neither to cause contraction of the model nor to affect the character of its contraction in response to ATP. They therefore do not act directly on the mechanochemical apparatus of the muscle. They probably act on membrane structures responsible for excitation of the living muscle, and only indirectly, in this way, on its contraction.

Among the substances investigated, with no effect on contraction of the fiber model, are acetylcholine, adrenalin, the acetylcholinesterase inhibitors eserine and diazopropylfluorophosphate, caffeine, neostigmine, strychnine, veratrine, ryanodine, digitoxin, histamine, quinine, and cocaine (Korey, 1950). However, caffeine, quinine, and digitoxin abolish the inhibitory action of the preparation of relaxation granules on ATPase contraction of the model. These substances are evidently able to depolarize the membranes of the sarcoplasmic reticulum, so that calcium ions can leave its cavities and activate and initiate the mechanochemical reaction; as a result, the fiber contracts.

Contraction of actomyosin threads obtained from contractile proteins of the uterus is not affected by estrogen, progesterone, oxytocin, acetylcholine, adrenalin, or histamine (Csapo, 1960).

I must emphasize here that it was through the use of these muscle models and the preparation of relaxing factor that the exact point of application for the action of these important pharmacological agents listed above on muscle was identified.

Contraction of a model fiber can be induced not only by ATP, but also by other nucleoside triphosphates and ADP. However, the concentration of ADP capable of causing the model to contract is higher than that of ATP by about one order of magnitude. The optimal ADT concentration, for instance, is 5×10^{-1} M, compared with 4×10^{-2} M for ATP. It is considered (Perry, 1957; Weber, 1958) that ADP cannot itself cause contraction of the model, but it is converted into ATP by the adenylate kinase present in the model: 2ADP → ATP + AMP.

The ATP thus formed reacts with the actomyosin system of the model fiber and a mechanochemical reaction takes place. AMP, like inorganic phosphates, is unable to cause contraction of the model.

Contraction of the glycerinated fiber has been described through the action of a totally unphysiological agent, namely, Nessler's reagent (HgI_2 + KI) (Laki and Bowen, 1955). KI, in the absence of HgI_2, also causes contraction of the model, but this is followed by rapid lengthening of the fiber. Laki and Bowen accordingly conclude that contraction is provoked by KI, and that the HgI_2 simply prevents the subsequent lengthening. Like KI, KCNS also causes contraction of the model fiber, followed by its lengthening. A mixture of KCNS and HgI_2 causes contraction without relaxation, so that in this case also HgI_2 plays the role of inhibitor of lengthening. Nessler's reagent can also cause plasticization of the model: after treatment with this reagent the model fiber can stretch twice or three times. The stretched fiber can contract again in Nessler's reagent. Nessler's reagent has the same action on actomyosin threads. Laki (1967) used the phase-contrast and electron microscopes to investigate changes taking place in the model fiber during its contraction under the influence of Nessler's reagent. He found that this contraction is completely indistinguishable from that evoked by ATP and consists of a decrease in size of the I-disks. This fact stands out on its own in the literature, where usually there are many statements to the contrary, namely to the high specificity of contraction of the muscle fiber, which can be evoked only by nucleoside triphosphates and ADP.

So far as we know at present, physical and chemical agents

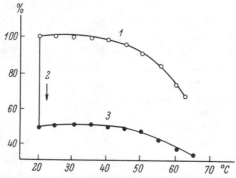

Fig. 10. Contraction of rabbit actomyosin thread by the action of ATP and thermal contracture of the same thread (after Portzehl, 1951). Abscissa, temperature; ordinate, length of thread under isotonic conditions (in % of initial length): 1) thread in medium without ATP, 2) contraction at 20°C by the action of 3×10^{-3} M ATP, 3) thermal contracture of thread already contracted in medium with 3×10^{-3} M ATP. Arrow indicates direction of change in length shown by curve 2 (shortening of thread).

causing muscle contracture do not cause contraction of the model. Heating to a high temperature (above 45-50°) shortens rabbit actomyosin threads (Portzehl, 1951), while heating to 50-55° or more leads to shortening of models of the rabbit muscle fiber (Ovsyanko, 1968; Suzdal'skaya and Troshina, 1968). This high-temperature contracture, however, has nothing in common with contraction of the models caused by ATP. It is observed not only in uncontracted threads, but also in threads contracted by the action of ATP (Fig. 10).

Laki's findings described above can be supplemented by the writer's own observations concerning the action of Nessler's reagent on models of ciliated epithelium. The possibility cannot be ruled out that these observations may lead to criticism of the interpretation of Laki's facts. As will be stated below (see page 107), the cilia of models of ciliated epithelium under the influence of ATP and Mg^{++} repeat the complete working cycle of flexion and extension over and over again, i.e., they flutter. However, after treatment with Nessler's reagent, they behave completely differently: each cilium shortens strongly and irreversibly. This type of contraction is not characteristic of cilia, and this suggests that

Nessler's reagent exerts some sort of physicochemical effect which, at least in the case of cilia, has no physiological meaning and is merely an artifact. This may also apply to the action of Nessler's reagent on the muscle model. In this case the effect of this reagent merely resembles the normal motile response of the fiber in its outward appearance: in both cases shortening of the contractile structures takes place.

This effect of KI and KCNS on the models may perhaps be based upon the ability of these substances (Szent-Györgyi, 1951) to depolymerize actin. Seravin (1967), who observed the wave-like motion of model flagella of Peranema trichophorum as a result of treatment with 0.3-0.6 M KI (or KCNS), also attributes this fact to depolymerization of the actin-like protein of the flagella.

Metabolic inhibitors, such as 1×10^{-2} M monoiodoacetate, 4.4×10^{-2} M sodium pyrophosphate, and 8×10^{-2} M mercuric tartrate and chloride, do not affect the contractility of the model. However, powerful sulfhydryl poisons, such as $1 \times 10^{-4} - 1 \times 10^{-3} M$ $HgCl_2$, 3×10^{-1} M H_2O_2, $5 \times 10^{-4} - 1 \times 10^{-3}$ M o-iodosobenzoate, and $1 \times 10^{-1} M$ mapharsen, inhibit it (Korey, 1950). In the case of o-iodosobenzoate, the inhibition can be abolished by rinsing the model for 2 h in a solution containing cysteine, and in the case of mapharsen, by rinsing it in salt solution for 10 min.

Contraction of muscle models is inseparably connected with hydrolysis of ATP. Experiments with models suggests that hydrolysis of ATP (and not merely its binding by contractile proteins) lies at the basis of contraction and provides it with its source of energy. Agents inhibiting ATP hydrolysis, but not preventing its binding (Salyrgan, for example) also act as inhibitors of contraction. The curves of ATP hydrolysis and intensity of model contraction as functions of temperature coincide: within the range 5-10°C, the value of Q_{10} for ATPase activity of rabbit actomyosin is 2.0. Q_{10} for maximal contraction (tension) of the rabbit fiber model is 1.8, and for its maximal shortening 1.9 (see also Fig. 11).

The values of the superoptimal ATP concentrations, i.e., the high concentrations which, on the one hand, inhibit the ATPase activity of actomyosin and, on the other hand, inhibit the contractility of the models obtained from the same object as the actomyosin, also coincide. The ATPase activity of actomyosin and the ability of models to contract are equally influenced by pH, ionic

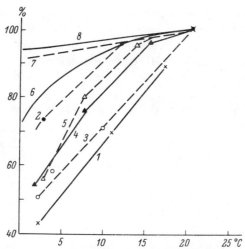

Fig. 11. Tension of various muscle preparations as functions of temperature. Abscissa, temperature; ordinate, tension developed by preparation (in % of tension at 20°C). 1, 2) Model of rabbit psoas muscle fiber (1, in 3×10^{-3} M ATP, 2, in 4×10^{-4} M ATP), 3) actomyosin thread from rabbit skeletal muscles in 3×10^{-3} M ATP, 4) model of fiber from smooth muscle of Anodonta in 2.5×10^{-3} M ATP, 5) actomyosin thread from smooth muscles of Anodonta in 1.5×10^{-3} M ATP, 6) skeletal muscle during tetanus, 7) resting skeletal muscle, 8) resting (without ATP) model of rabbit psoas muscle fiber. Figure taken from Portzehl (1951) with the addition of curve 4 (Ulbrecht and Ulbrecht, 1952) and curve 5 (Dorr and Portzehl, 1954).

strength, and Mg^{++} concentration. Calcium ions are strictly essential for both processes, although in a very low concentration, of the order of 1×10^{-6} g-ion/liter.

Relaxation of the model fiber takes place under conditions enabling ATP to be bound by contractile proteins but, at the same time, preventing its hydrolysis. Binding of ATP plasticizes the models: bridges from the subfragments-1 of heavy meromyosin are separated from the actin protofibrils, the thick and thin protofibrils are disconnected, and contraction does not take place because there is no hydrolysis of ATP. As a result, the fiber lengthens under the influence of the applied load.

If no ATP is present in the medium, neither relaxing factor nor Salyrgan can itself evoke relaxation of the model. Their role is simply to inhibit hydrolysis of ATP, and in that way to inhibit contraction of the model; under those conditions the plasticizing action of ATP is manifested as relaxation of the model. Relaxation can also be caused by adding, instead of nucleoside triphosphate, other organic or inorganic polyphosphates, such as sodium pyrophosphate, to the medium because all these substances will plasticize the model fiber (Fig. 7).

Spontaneous relaxation of the model fiber has also been described. A glycerinated fiber which has contracted under the influence of ATP will relax if the medium in which it was placed and in which the ATP acted is not buffered (Ranney, 1954). Relaxation of the fiber is evidently influenced by the ATP—creatine phosphate system (Goodall and Szent-Györgyi, 1953). Creatine phosphate rephosphorylates ADT to form ATP, the complex of myosin with actin dissociates, and the fiber relaxes. Later, with the aid of creatine phosphate (CP), which he added to the medium with ATP ($4 \times 10^{-3} M$ ATP, $1 \times 10^{-1} M$ CP, $4 \times 10^{-3} M$ MgCl$_2$, pH 7.0), Goodall (1956) obtained oscillation of a model of the rabbit psoas muscle fiber. This oscillation took place, in Goodall's opinion, through the combined action of ATP (including contraction) and CP (inducing relaxation).

The models have shed light on the nature of the rhythmic activity of insect flying muscles. These muscles are characterized by high-frequency oscillatory movements at a rate of the order of 100-1000 contractions per second. Whereas in the frog sartorius muscle, for example, the ratio between the number of electrical stimuli and mechanical responses is 1:1, in the insect flying muscles one spike corresponds to many contractions of the fibers. In the blowfly wing, for instance, action potentials have been recorded at a frequency of about 3/sec, while the wing itself flaps at a frequency of about 120/sec (Hanson and Lowy, 1960). This suggests that in the case of the flying muscles of insects, rhythmic activity is an intrinsic property of the contractile system itself, and not of the excitation coupled with contraction. This problem was investigated with the aid of models possessing a contractile system but devoid of excitation processes and, evidently, of structures capable of excitation. These investigations (Jewell et al., 1964; Jewell and Rüegg, 1966; Pringle, 1967, 1968) showed that the work of myofibril models prepared from the flying muscles of insects is oscillatory in character. Consequently, the rhythmic mode of working, and the continuity of contradiction and relaxation in these muscles are properties of the contractile system itself and they are not assigned by some form of pacemaker lying outside the motile system. The mechanically active elements of these muscles can probably respond to a change in their own length by a change in the opposite direction: to lengthening by shortening and to shortening by lengthening. As a result, rhythmic movements are produced. This evidently is the explanation of the rhythmic motor activity of flagella and cilia in which, as we shall see below, it is also an intrinsic property of the motile system.

Models from flying muscles of insects also oscillate in the presence of ATP, Mg^{++}, and Ca-EGTA even if treated after glycero-extraction with the detergent Tween-80, which dissolves membranous structures so that remnants of noncontractile proteins and also the contractile proteins of mitochondria and other membranous structures are extracted from the fibers. Such models are the pure isolated motile skeleton of the fiber (Abbott and Chaplain, 1966). A solution containing EGTA as well as KCl and Mg^{++} (i.e., calcium-free) sends them into contracture (rigor). A solution containing KCl, ATP, Mg^{++}, and EGTA plasticizes and relaxes them; this corresponds to the state of the living muscle at rest. In a solution with $(ATP-Mg_2)^{--}$ and an equilibrium concentration of Ca^{++} (between 9×10^{-8} and 1×10^{-5} g-ion/liter), which is produced with the aid of Ca-EGTA buffer, they oscillate. With an increase in the Ca^{++} concentration (within the limits specified) the amplitude of oscillations of the model increases, and in some objects so also does the force of contraction. If the oscillating model is transferred to calcium-free solution its tension falls and the oscillations disappear. If it is returned to the solution containing Ca^{++}, the oscillations reappear. By changing the Ca^{++} concentration in the medium, the investigator can thus switch on and off, strengthen and weaken the oscillatory movements of this model. In an appropriate solution the model will continue to perform oscillatory activity for many hours (the longest period so far recorded is 23 h). The amplitude of the oscillations increases up to $[Ca^{++}] = 5 \times 10^{-7}$ g-ion/liter, but the oscillation itself does not disappear even if $[Ca^{++}]$ is increased up to 1×10^{-5} g-ion/liter (the upper limit of possible equilibration of the solution with EGTA).

---

*See also the section on preparation of models of wing muscles of insects and their myofibrils in Chapter IV.

The work of this model is disturbed by Mg^{++} ions. Magnesium in a solution inducing oscillation must be in the $(ATP-Mg_2)^{--}$ form (Jewell and Rüegg, 1966).

Glycerinated models from heart muscle respond to ATP only by contraction, and they do not reproduce the complete rhythmic working cycle of the myocardial tissue. The rhythmic activity of the heart is not myogenic but membranous in nature; it is regulated via the sarcoplasmic reticulum by impulses generated by the specific (noncontractile) tissue of the pacemaker.

The fundamental tendency in modern cytology, biochemistry, and molecular biology is to combine the structural and functional approaches to the investigation of submicroscopic structures. Muscle models were and still are an excellent object for putting this tendency into practice, especially as regards the study of the mechanism of protoplasmic contractility. By means of these models the distribution of the concrete contractile proteins in the sarcomere has been determined, changes taking place in the sarcomere during contraction (behavior of the protofibrils, structural modifications to the bonds between them) have been studied, and attempts have been made to elucidate the mechanism of conversion of chemical energy into mechanical (Hanson and Huxley, 1955; Hanson and Lowy, 1960; Huxley and Hanson, 1960; Aronson, 1965; Shtrankfel'd et al., 1966; Perry, 1967; Chaplain, 1970; see also: Aspects of Cell Motility, 1968). The submicroscopic changes leading to contraction of the sarcomere or smooth muscle fiber have been studied in living muscles and in models by the x-ray diffraction method at low and medium angles (Huxley and Brown, 1967; Huxley, 1968, 1969; Pringle, 1968).

It is by the use of models that the fluorescent antibody technique has been successfully used to detect individual contractile proteins in muscle fibers and other cells (Aronson, 1965; Chaplain, 1970) and even to localize the relaxation system in cells and its translocations (Kinoshita and Yakazi, 1967; Kinoshita, 1968). In investigations of this type, an antibody is produced against the particular protein or protein system under study, the antibody is labeled with a fluorochrome, the model of the test object is then placed in a medium containing the fluorescent antibody, and the antibody diffuses into the model (it could not penetrate so easily into a living cell) and reveals the protein antigen, thus solving the problem.

The x-ray diffraction method at low and medium angles, which reveals periodic structures and changes in their period under certain experimental conditions, is particularly valuable because it can be used both with models and with intact living muscles and cells. Besides its other uses, this means that the x-ray diffraction method can be used to verify the results obtained with models by information obtained from living objects.

Reedy and co-workers (1965), for instance, used as their test object glycerol-extracted fibers of insect flying muscles, which are in a state of rigor in a solution containing EGTA and $MgCl_2$, and in a relaxed state in the same solution after the addition of ATP. They studied these models under the electron microscope and also by the low-angle x-ray diffraction method and compared the character of the links between the protofibrils of the muscle in relaxed and contracted states. They found that bridges crossing from myosin to actin fragments in the resting fiber are arranged at right angles to both, and that they are not connected to the actin fragments. In the contracted myofibril, on the other hand, the bridges are attached to the actin filaments, they are arranged at an acute angle to the protofibrils, and the "arrows" thus formed have their pointed ends facing the center of the sarcomere (Fig. 4; for a schematic representation, see Fig. 12).

These observations, in the writer's view, are of great potential value in that they open the road to similar investigations using different experimental conditions of contraction, and their great importance evidently lies in the fact that they reveal the im-

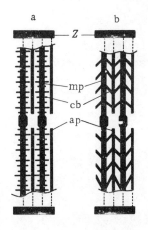

Fig. 12. Scheme of relations between myosin and actin filaments and connections between them by means of cross-bridges in relaxed (a) and contracted (b) glycerol-extracted myofibrils of the dorsal longitudinal muscle of Lethocerus maximus (from Reedy et al., 1965). The reason for the difference in lengths of the bridges in (a) is that the projection of the bridges on the plane of the diagram is shown, and bridges surrounding the myosin protofibril are stretched in different azimuthal directions. Relaxation, in solution containing 5×10^{-2} M KCl, 5×10^{-2} M $MgCl_2$, 5×10^{-3} M ATP, 4×10^{-3} M EGTA (e.g., without Ca^{++}); rigor in solutions without ATP and without Ca^{++}; KCl, $MgCl_2$, pH the same as before. mp) myosin, ap) actin protofibrils, cb) cross bridges; Z-membranes are identified.

portant role of the bridges in the contraction mechanism. Pringle (1968) considers that calcium ions settle on the cross bridges and strengthen interaction between the actin and myosin fragments by increasing the number of links between them. This leads both to the development of tension and to activation of relaxation by ATP. A quantitative correlation exists between these two processes: the intensity of one of them is directly proportional to the intensity of the other, and at the same time, to the number of bridges. Tension develops as a result of the bending movements of the bridges which, through hydrolysis of ATP, propel the actin protofibrils along the myosin and generate the sliding movement of one relative to the other, thereby approximating neighboring Z-membranes (Huxley, 1969).

It is interesting to note that bridges belonging to one myosin protofibril do not bend synchronously, but regularly after each other in phase (Huxley and Brown, 1967; Huxley, 1968). This rhythm is described as metachronal. A wave of bending arises which resembles the wave during oscillation of the cilia of ciliated epithelium or of paramecium. The similarity between bending and straightening of the bridges in the oscillating myofibril of the insect flying muscle is evidently greater still. Whether this similarity is purely outward, or reflects some related or common origin of these forms of protoplasmic movement cannot yet be answered because the exact mechanism of bending and straightening of the cilium is not yet known. The investigation of the function of the bridges, which can now be isolated preparatively (Huxley, 1969) and of the "arms" of ciliary filaments (Gibbons, 1967), which are probably analogous in their function to the bridges, is the next task which must be undertaken. The study of these structures or suborganoids will not only shed light on the mechanism of working of the muscle and cilium (flagellum), but will also provide evidence concerning the genesis and evolution of motility.

Models of Contraction of Nonmuscle Cells and Their Organoids

The preparation and study of models of muscular contraction have opened the way to the study of the mechanism of nonmuscular forms of contraction.

Hoffmann-Berling (1953), a collaborator of Weber, treated tissue cultures of fibroblasts in the same way as muscle fibers are treated for the preparation of models. He dound that the cells die but their insoluble structures remain intact. The cells are actually dead: their cytoplasm is coagulated; on staining with acridine orange they give red fluorescence, and not green like living cytoplasm; the cells are easily digested by trypsin. If, after glycerol extraction, the cells are gradually taken into a glycerol-free saline solution, the shape of the cells and their nuclei, nucleoli, chromosomes, and division spindles remains unchanged; only the mitochondria become round. In this way a cell model similar to the model of the muscle fiber is obtained (Fig. 13). If

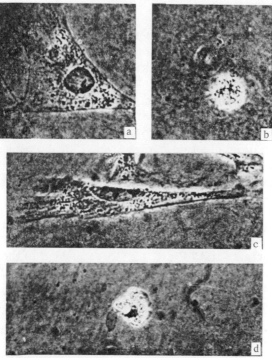

Fig. 13. Glycerinated models of chick amniotic fibroblasts contracting under the influence of ATP (from Hoffmann-Berling, 1954a). (a) and (c) without ATP; (b) and (d) 12 min after immersion in solution containing $2 \times 10^{-3}\,M$ ATP; $\mu = 0.14$, pH 7.0, t = 37°C.

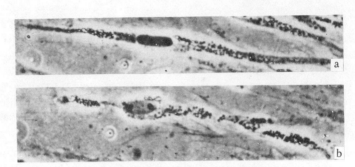

Fig. 14. Glycerinated model of chick scleroblast contracting isometrically in viscous medium not permitting it to shorten under the influence of ATP (Hoffmann-Berling, 1954a): (a) without ATP; (b) 12 min after immersion in solution containing 8×10^{-3} M ATP; $\mu = 0.17$, pH 7.2, t = 37°C. Force of contraction disrupts internal structure of the model.

such a model, extracted for several days and then washed to remove the extracting solution, is treated with ATP and Mg^{++} in certain concentrations, the cells as they contract become spherical in shape (Hoffmann-Berling, 1953, 1954a). The model cells thus become similar to living cells during prophase or during exposure to stimulation. The cytoplasm of living connective-tissue cells is known to contract in response to mechanical stimulation, or during the action of an electric current, ethyl alcohol, procaine, $CaCl_2$, KCl, trypan blue, India ink, and adrenalin, and the cells become rounded (Vol'fenzon, 1954; Aleksandrov and Vol'fenzon, 1956).

On placing these cell models in a viscous medium preventing them from shortening Hoffmann-Berling observed contraction of the cell contents if ATP and Mg^{++} were added to the medium. The outlines of the cell remained unchanged because of the high viscosity of the medium, but the cytoplasm broke up into parts because of the pull exerted by its contractile elements (Fig. 14).

ATP is specific in its action: none of the inorganic or organic phosphates tested (including AMP and CP) can give the same effect. ATP can be replaced only by other nucleoside triphosphates.

A number of cell models reproducing other forms of protoplasmic movement have recently been obtained. However, it was concluded from the first of these models described above that contractility of nonmuscular cells and of the different types of muscles is based on the same biochemical process: the reaction between contractile protein structures of ATP (Hoffmann-Berling and Weber, 1953; Portzehl, 1954).

Mechanisms of relaxation in muscle and nonmuscle models also are similar. The relaxing granules which were mentioned in the section on contraction of muscle models also act on models of fibroblasts (Kinoshita et al., 1964). Just as with muscle models, relaxing granules prevent cell models from contracting. These authors first studied the action of relaxing granules derived from muscles on models of fibroblasts and obtained the following results. The presence of muscle granules in the medium surrounding models of chick embryonic fibroblasts in a concentration of 0.5 mg granule protein/ml medium almost completely abolishes the contractile response of the models to ATP and Mg^{++}. However, if the granule preparation is first treated with 1×10^{-4} M $CaCl_2$, these granules can no longer inhibit contraction of the models. The granules are evidently saturated with calcium ions and can no longer take them from the medium surrounding the model; calcium ions can now reach the contractile elements of the model and activate their reaction with ATP. Kinoshita and co-workers succeeded in isolating relaxing granules from nonmuscle cells also, in fact from 48-h cultures of chick embryonic connective tissue. Purified preparations of these cell granules inhibit contraction of model fibroblasts in precisely the same way as preparations of muscle granules. Their activity can also be prevented by calcium ions.

Similarity between the reaction of muscle and cell models to relaxing granules is also revealed in experiments to study the effect of certain poisons, notably the organic mercury compound Salyrgan (mersalyl) on these granules. If relaxing granules are treated with Salyrgan, they will no longer act on muscle models, and this suppression of the activity of the granules is irreversible. However, if the granules are kept in ATP solution and then treated with Salyrgan, unlike in the preceding case, their activity can be restored by cysteine. This is probably a manifestation of that widespread phenomenon – protection of an enzyme by its substrate (in this case, ATP).

The situation is exactly the same if the experiment to study the action of relaxing granules is performed, not on models of the muscle fiber, but on models of fibroblasts. It does not make the slightest difference whether the granules are obtained from muscles or from connective-tissue cells.

It can be concluded from these findings that the mechanism of relaxation in muscle fibers and nonmuscle cells also is probably identical.

Models of fibroblasts have provided direct proof that different types of contractility are based on the same principle of energetics. This conclusion was drawn from a comparison of the properties of muscle fiber and fibroblast models. These two types of models are similar in the following respects.

1. Both types of models — muscle and cell — contract in the presence of ATP. The action of ATP in this respect is specified and it can be replaced (although less successfully), only by other nucleoside triphosphates. Rinsing out the ATP stops the contraction unless it has reached its maximum.

2. Cell and muscle models can contract up to a maximum of 20% of their initial length.

3. Plots of contraction amplitude versus ATP concentration (and also versus ATP concentration) for models of myofibrils and for cell models coincide (Fig. 15, compare with Fig. 8).

The character of the plot of contraction amplitude versus ATP concentration is identical for all models: with an increase in concentration the contraction at first increases, reaches an optimum, then decreases and eventually disappears. The optimal concentration is lower for models from smooth muscles and higher

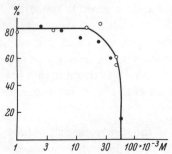

Fig. 15. Contraction of models of chick amniotic fibroblasts as a function of ATP (circles) or ITP (dots) concentration (from Hoffmann-Berling, 1954a). Abscissa, ATP or ITP concentration; ordinate, degree of shortening of models (in % of initial length). Extraction for 39 days. Reactivating solution: ionic strength 0.15 for nucleoside triphosphate concentration up to 1×10^{-2}, up to 0.35 for higher concentrations; pH 7.2; Mg^{++} 5×10^{-3} g-ion/liter in experiments with ATP, and 1.5×10^{-2} g-ion/liter in experiments with ITP.

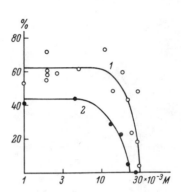

Fig. 16. Contraction of models of chick scleral fibroblasts as a function of temperature (from Hoffmann-Berling, 1954a). Abscissa, ATP concentration; ordinate, degree of shortening of models (in % of initial length). Extraction for 8 days. 1) Contraction at 37°C; 2) at 4°C.

for models from skeletal muscles and fibroblasts. ATP concentrations too high to cause models to contract are known as superoptimal.*

4. Both models contract more slowly and less completely if the temperature is lowered (Fig. 16; compare with Fig. 11). The optimal ATP concentration falls as the temperature is lowered from 20 to 0°C.

5. The optimal pH of the medium for all models is about 7.0. At pH values below 5.6 and above 8.0 no contraction can be obtained (Fig. 17). The ionic strength of the medium must be less than 0.15.

6. Both muscle and cell models function only in the presence of magnesium ions, the concentration of which must be between 1×10^{-4} and 1×10^{-3} M in the case of ATP and 10-100 times high-

---

*In the writer's experience the action of superoptimal concentrations on cell models (in this case, models of frog ciliated epithelium, which will be described later) is irreversible: models treated with ATP is superoptimal concentrations are subsequently incapable of contracting in the presence of optimal ATP concentrations. This implies that ATP, in superoptimal concentrations, damages cell models. If this is true, it is highly probable that this action of ATP is not connected with the high-energy end of the molecule, and would be reproduced if the ATP were replaced by AMP. Experiments in which AMP was used in concentrations corresponding to superoptimal ATP concentrations could confirm, or otherwise, the hypothesis that the superoptimal effect is due to injury to the contractile structures. The opinion has also been expressed, no doubt on the basis of weightier evidence, that the superoptimal effect is fundamentally a manifestation of enzyme-substrate inhibition (Hoffmann-Berling, (1959)). One explanation, of course, need not necessarily exclude the other.

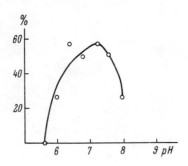

Fig. 17. Contraction of models of chick scleral fibroblasts as a function of pH (from Hoffmann-Berling, 1954a). Abscissa, pH; ordinate, degree of shortening of models (in % of initial length). Extraction for 10 days. Reactivating solution: 2.5×10^{-3} M ATP, $\mu = 0.13$, $t = 37°C$.

er if ITP is used (Fig. 18). If the Mg^{++} concentration in the medium is low, the maximal amplitude of the contraction, the rate of shortening, the maximal tension, and the rate of its development all tend toward zero.

7. The work of both types of models is suppressed by the same inhibitors. Their contraction under the influence of ATP and Mg^{++} is reversibly inhibited also by poisons containing heavy metals (such as Salyrgan) and by organic polysulfo-acids (Oxarsan,* Fouadin, Germanin). The same general principle applies to models of both types: if the concentration of poison is such that it depresses the contractility of the models incompletely, as a rule the degree of depression rises with an increase in ATP concentration. Monoiodoacetate has no action on models.

8. In the case of incomplete inhibition of both models, the faster the rate of ATP hydrolysis, the slower the rate of contraction of the model.

9. ATP causes not only contraction, but also "softening" or plasticization of boyh types of models. The rounding of cell models, like muscle models, is not only due to the plasticizing action of ATP; it is a true contraction, for in superoptimal ATP concentrations plasticization is present but contraction is absent.

10. If extraction is of short duration (several hours) the contractility of both types of model is reduced, but it can be activated to the normal level by addition of Ca^{++}. Contractility of a model extracted for a longer time (in optimal ATP concentrations) is not increased by calcium ions. The reduced contractility of

---

* m-Amino-p-hydroxyphenylarsene (III) oxide.

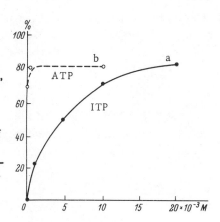

Fig. 18. Contraction of models of chick amniotic fibroblasts as a function of concentration of Mg^{++} and Ca^{++} ions (from Hoffmann-Berling, 1954a). Abscissa, Mg^{++} concentration; ordinate, degree of shortening of models (in % of initial length). Points represent: a) 2.5×10^{-3} M ITP without addition of Mg^{++} and Ca^{++}, and also on addition of 2×10^{-2} g-ion/liter Ca^{++}, but without Mg^{++}; b) 2.5×10^{-3} M ATP without addition of Mg^{++} and Ca^{++}, extraction for 32 days. Reactivating solution: $\mu = 0.15$, pH 7.2, t = 37°C.

briefly extracted models, as experiments on muscle models showed originally, is due to the fact that the relaxation system which prevents contraction by binding calcium ions has not been disrupted in them. Granular relaxing factor has been prepared both from muscles and from fibroblasts, and the properties of the relaxing granules obtained from both sources were found to be very similar.

Both activation and inactivation of muscle models are thus produced by the same conditions as the corresponding processes of cell models.

11. Contraction of models of fibroblasts develops as a result of active tension just as in muscle models. This has been shown by direct experiments (Hoffmann-Berling, 1956). Total models were prepared from pieces of tissue culture and tested on a strain gage in the presence of ATP. They were found to develop a tension of 22 g-wt/cm^2, compared with a tension of 4 kg-wt/cm^2 for muscle models. Hoffmann-Berling attributes this difference to the great difference between the content of contractile protein in muscles and in connective-tissue cells (about 100 times greater in the former than in the latter), and also to the fact that the cells are freely linked together so that during contraction they slide along each other, and the total tension in the piece of tissue as recorded by the strain gage is thus smaller than the sum of the tensions of the individual cells.

The following comment must be added to this explanation. In a muscle fiber (and also in the model) the contractile structures are oriented longitudinally, and for this reason tension of each element also develops in this same direction; the resultant, directed along the fiber, is equal to the arithmetical sum of the tensions of all its elements. In fibroblasts, on the other hand, the mechanically active elements are oriented in different directions. Tension along the length of the piece of tissue must be substantially less than the sum of the contractions of the contractile structural elements. (For a discussion of the ultrastructural composition of fibroblast models and of models of some other cell types, see pages 65-69.)

Hoffmann-Berling (1956) isolated a contractile protein from the cells of Jensen's and Yoshida's rat sarcomas. It is similar to actomyosin in extractability, in the character of the change in its viscosity on contact with ATP, and in its ATPase activity, which can be activated and inhibited to an equal degree. Its aggregates shrink and form a superprecipitate on contact with ATP. It differs only in the rate of ATP hydrolysis, which is 100 times slower in the sarcoma protein. However, the rate of hydrolysis is equally dependent on temperature, Mg^{++} concentration, and ionic strength as with actomyosin. An actomyosin-like protein, known as thrombostenin, has been obtained from platelets and filaments have been prepared from it which contract under the influence of ATP and Mg^{++} (Bettex-Galland and Lüscher, 1961; Bettex-Galland et al., 1962).*

The only significant difference between the properties of muscle and cell glycerol-extracted models is in the rate of their contraction: at 20°C the muscle model contracts at a velocity corresponding to 200% of its initial length per second while the cell model contracts at a velocity of 0.50% of its initial length per second. In other words, the muscle model contracts 400 times quicker than the cell model. This is explained by the lower ATPase activity of the contractile protein of the cell. The temperature coefficient Q_{10} of the contraction velocity is ~2 for both muscle and cell models.

It can be concluded from these observations that the biochemical nature of muscle and cell contraction is qualitatively very similar. There is every reason to suppose that muscle con-

---

*See Notes Added in Proof, page 169.

tractility has developed in evolution from the contractility which is an inherent property of protoplasm in general.

Cell models becoming rounded in the presence of ATP and Mg^{++} were obtained by Hoffmann-Berling not only from connective-tissue cells, but also from epithelial and ascites cells. In addition, an analogy can be drawn between muscle contraction and the rounding of amebas. Rounding of living amebas takes place through the action of chemical, electrical, mechanical, and, probably, other physical agents. The technique of model preparation can also be used with amebas, models of which contract when brought into contact with ATP and Mg^{++}. Models of amebas obtained by glycerol extraction are mentioned in one paper by Hoffmann-Berling (1954). He wrote, in particular, that contraction of ameba models induced by ATP is so strong that the pseudopodia, which usually stick to the slide, very often break away from the body of the model ameba on the addition of ATP. Detailed descriptions of models of amebas and the techniques used to prepare them have not been published by Hoffmann-Berling. Simard-Duquesne and Couillard (1962a) originally tried to obtain models of Amoeba proteus by using Hofmann-Berling's usual technique but were unsuccessful. They modified this technique by first removing Ca^{++} from the amebas with versene solution, reducing the ionic strength of the extracting and working solutions, and by carrying the objects in stages through solutions with gradually increasing concentrations of glycerol, and thereby succeeded in obtaining good models of amebas. The models which they obtained differ from the models of amebas described by Hoffmann-Berling: their pseudopodia are retracted during extraction. In the ameba models obtained by Simard-Duquesne and Couillard, the cell body itself contracts on the addition of ATP (Fig. 19).

The features of similarity between ameba models and those obtained from muscle and other cell types are as follows. ATP specifically induces contraction. The presence of Mg^{++} is an essential condition for contraction (its optimal concentration is 3×10^{-3} g-ion/liter). Sulfhydryl poisons (p-chloromercuribenzoic and p-chloromercuriphenylsulfonic acids, 1×10^{-4} M) inhibit contraction of the models but their removal from the models by rinsing with 0.02 M cysteine restores the ability of the models to respond by contraction to 5×10^{-4} M ATP and 3×10^{-3} M $MgCl_2$; the pH of the extracting and working solutions must be about 7.0.

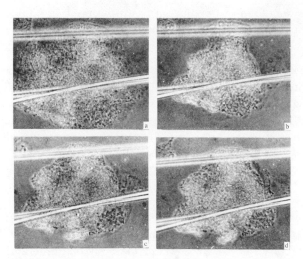

Fig. 19. Glycerinated model of Amoeba proteus (from Simard-Duquesne and Couillard, 1962a). (a) Before transfer to solution with ATP; (b) 1 min after transfer to solution containing 5×10^{-4} M ATP and 3×10^{-3} M MgCl$_2$ — contracted model; (c) 5 min after transfer to solution containing 5×10^{-3} M ATP (concentration 10 times greater than that causing contraction in b) and 3×10^{-3} M MgCl$_2$ — stretched model; (d) 5 min after transfer back to solution containing 5×10^{-4} M ATP and 3×10^{-3} M MgCl$_2$ (the same concentrations as in b) — repeated but extremely weak contraction.

Ameba models differ from other cell models in the instantaneous character of their contraction. This brings them more into line with muscle models. A feature which distinguishes these models from all others is that they will contract weakly in the presence of 3×10^{-5} M MgCl$_2$ without ATP. However, their powerful and rapid contraction requires the presence of ATP in the medium. In this case it can be assumed that there are two different mechanisms of contraction, one of which requires both ATP and Mg^{++} and resembles muscle contraction, while the other is triggered by Mg^{++} alone and resembles that observed in Vorticella (see pages 90-95) except that in Vorticella the role of Mg^{++} is played by Ca^{++} ions.

Finally, another interesting phenomenon is observed in model amebas: if the ATP concentration in the activating solution is 5 ×

10^{-4} M or 1×10^{-3} M, sometimes after contraction slight relaxation is observed, but if this concentration is higher (5×10^{-3} M) the contraction is followed by a powerful and rapid relaxation. A small contraction can be induced after this relaxation by means of weak concentrations of ATP (6×10^{-4} M).

It can be concluded that ATP, in a low concentration, is an agent inducing contraction. In high concentration, on the other hand, such as arises by diffusion of ATP into a model, it is an agent of relaxation.

Seravin (1967) observed a contraction in model amebas which differs from that described above. Under the influence of 1×10^{-3} M ATP and 1×10^{-3} M Mg^{++} the cytoplasm of the models separates from the plasmalemma and condenses into a compact mass, although the plasmalemma retains its original irregular outlines (Fig. 20). During this contraction the tiny granules which float between the condensed cytoplasm and the plasmalemma can be seen to move toward the plasmalemma. They are evidently moved by the current of water displaced during contraction of their cytoplasm. In that case, the isodimensional contraction of the amebas can be attributed to the syneresis of their cytoplasmic proteins, by analogy with the syneresis of the actomyosin aggregates under the influence of ATP.

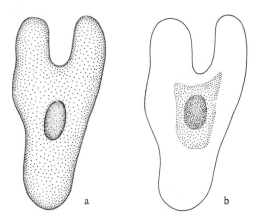

Fig. 20. Diagram showing contraction of model of Amoeba proteus under the influence of ATP (from Seravin, 1967): a) before immersion in solution containing ATP, b) in solution containing ATP.

Simard-Duquesne and Couillard (1962) extracted a protein resembling myosin ATPase from amebas and purified it. This is a further piece of evidence in support of the similarity of the contraction mechanisms in amebas and muscles.

A preparation of a killed ameba in which the cytoplasm moves, and this movement is dependent on ATP, has recently been obtained (Gicquaud and Couillard, 1970). These workers placed amebas in an iso-osmotic salt solution of the following composition: 1 mMATP, 2 mMMgCl$_2$, 0.75 M EGTA, 25 mMKCl, 2.8 mM NaCl. In this solution the plasma membrane of the ameba bursts. The cytoplasm flows out through the teat in the membrane but does not spread, and simply remains as a single compact mass. The α-granules of the cytoplasm begin to move about, and in some areas a violent movement of the cytoplasm takes place. After 2-15 min the cytoplasm coagulates, and it ceases to move. Movement in the preparation just described takes place only in a medium containing Mg^{++} and EGTA. Removal of the Mg^{++} by means of 5 mM EDTA completely prevents movement, and a fiber of indefinite length and up to 1 μ in diameter appears in the cytoplasm. The optimal EGTA concentration in the medium is 0.2-1.5 mM and the optimal ATP concentration 0.5-3.0 mM. If no ATP is added to the medium, either no movement takes place or it is slow.

Electron-microscopic techniques have been extensively used in studies on glycerol-extracted models derived from various types of cells. This method has been employed by Schäfer-Danneel and Weissenfels (1969) in their investigations on the models of fibroblasts (myoblasts) maintained in a tissue culture. Extraction was found to remove a large part of the ground substance of the cytoplasm and a certain proportion of the membrane structures from the fibroblasts. Remnants of cristae and mitochondria and many vacuoles could be seen, and the centriole was well preserved. Two types of fibrils, thick about 80 Å in diameter and thin from 20 to 50 Å in diameter, could be distinguished more clearly in the models than in preparations from living cells. The thick and thin fibrils are interwoven and are located chiefly at the periphery in both models and cells. After treatment of the model with ATP solution containing Mg^{++}, the models contract. The ectoplasm condenses around the nucleus and draws away from the cell membrane and outer layer of coagulated cytoplasm; that is why during contraction of fibroblast models their outlines sometimes remain unchanged. The fibrillary network condenses.

Electron-microscopic studies of glycerol-extracted cell models have also been undertaken on other objects, or on similar material but by other investigators. The pictures observed are generally analogous and differences have been found only in the size of the fibrils in different objects. Keyserlingk and Schwarz (1968) emphasize that models of fibroblasts have no cytoplasmic membrane and that their other membrane structures are poorly preserved; they point out that the thin (20-40 Å in diameter) fibrils are longer than the thick fibrils (diameter 100 Å), that both are oriented along the projecting processes of the model, and that during ATP-induced contraction they concentrate toward the center. The same fibrils are found in preparations from living polymorphonuclear leukocytes (Keyserlingk, 1968). Similar fibrils have been found in glycerol-extracted models of connective-tissue cells from the tunica propria of the frog small intestine (Beck et al., 1969). Fibrils of two types, thin (40-100 Å in diameter) and shorter, fusiform, thick fibrils (160-200 Å in diameter), have also been found in models of amebas (Schäfer-Danneel, 1967). Both types interweave in the living ameba to form a network which is uniformly distributed throughout the cell body, while in the model they are gathered into bundles. ATP causes condensation of this network, and this is accompanied by slight contraction of the body of the model ameba. It will be noted that Schäfer-Dannell was unable to prevent contraction of the cell during extraction, so that some redistribution of the fibrils took place during preparation of the model and contraction of the model in response to ATP was slight. Pollard et al. (1970) described the results of electron-microscopic investigations of glycerol-extracted models of Acantamoeba casellanii and similar results obtained by other workers studying Nitella cell fibroblasts and epidermal cells of Ascidia. Similar results are described by Ishikawa et al. (1969), who studied models of fibroblasts, chondrocytes, and epidermal and nerve cells. All the results were consistent, and in every case contractile fibrils were found. Methods of isolating these fibrils from amebas (Morgan et al., 1967) and from chondrogenic cells (Ishikawa et al., 1969) have been suggested.

Fibrils have also been found in the plasmodia of the myxomycete Physarum polycephalum and also in their models (Kamiya, 1968; Wohfarth-Bottermann, 1968). The diameter of the thin fibrils is 50-70 Å. Cytochemical studies have shown that they hydrolyze ATP. The actomyosin-like protein obtained from the plasmodia of this same myxomycete, dissolved in 0.6 M KCl, has been investigated electron-microscopically (Nachmias and Huxley, 1970). Bead-like filaments 0.1-1.2 μ in length and about 50 Å in diameter were found, sometimes with arrow-shaped structures attached to them, and occasionally connected, in turn, with thin bristles a few hundred Angström units in length. After treatment of the protein with 0.5-5.0 mM ATP the arrow-shaped structures are detached from the bead-like filaments, and the latter become unbranched and similar to filaments of muscle F-actin. Filaments of the same type are also formed during repolymerization of plasmodial actin: there is a complete analogy in this case with artificial filaments of muscle actin.

The use of models has thus shown that thin fibrils of non-muscle cells are similar to the actin protofibrils of muscles. The latter, like artificial actin filaments, are able to join with subfragments-1 of H-meromyosin, thus forming the characteristic arrow-shaped structures. H-meromyosin can therefore be regarded as a specific reagent for the selective detection of actin. On addition of H-meromyosin to the medium containing glycerol-extracted models of chondrocytes, fibroblasts, and epidermal cells H-meromyosin penetrates into the models (such experiments, incidentally, can be carried out only on models and not on living cells — yet another advantage of models) and actively joins the thin fibrils. The arrow-shaped structures which are formed are evidence of the actin nature of the thin fibrils of nonmuscle cells. Neither collagen fibrils, nor tonofibrils, nor microtubules, nor 100-Å filaments (thick fibrils), nor the various membranes join with H-meromyosin. The same linkages with H-meromyosin are also formed in vitro by thin filaments isolated from living chondrocytes.

Similar results were obtained on models of amebas (Pollard et al., 1970) and on thin fibrils isolated from Physarum polycephalum, and also on artifical filaments resynthesized from the actin-like protein of plasmodia (Nachmias and Huxley, 1970).*

---

*See Notes Added in Proof, page 170.

Hence, in all the cases mentioned above the thin filaments were shown to have the nature of actin. It is only necessary to mention the actin-like nature of the contractile protein of the phage tail sheath (Poglazov, 1966) to demonstrate clearly that actin is the most widely spread contractile protein. Experiments similar to those just described must now evidently be carried out (which is quite possible) on models of Acetabularia, on trichocysts, on dynein (the contractile protein of cilia),* and on flagellin (the contractile protein of bacterial flagella). The evolution of the mechanochemical function of protein and its place at the basis of the appearance and evolution of the sliding filament mechanism are interesting problems which are now just beginning to be studied.

Loewi (1952) isolated an actomyosin-like protein from the plasmodium of the myxomycete Physarum polycephalum. Several workers subsequently studied the contractile proteins of myxomycetes in detail (Ts'o et al., 1956a,b, 1957; Nakahima, 1960; Hatano and Oosava, 1966; Hatano and Tazava, 1968) and, as a result, found that in many respects their properties were similar to those of muscle contractile proteins, from which they concluded that they are responsible for movement of the plasmodial cytoplasm.

Movement of the cytoplasm in plasmodia of myxomycetes has been and is being studied intensively by Japanese workers headed by Kamiya, who has alluded to the tempting prospects of using the technique of glycerinated models with this particular test object and to the hopes of reproducing ATP-induced currents of protoplasm in such models (Kamiya, 1959).

The first results along these lines were published by Ohta in 1960 (cited by Kamiya and Kuroda, 1965). Ohta cut glycerinated plasmodia into strips which, when placed in medium containing ATP and Ca^{++} or Mg^{++} contracted, although only by 4%.

Kamiya and Kuroda (1965) tackled the experimental preparation of models from plasmodia of myxomycetes in a different way. By using highly ingeneous experimental techniques, these workers were able to place a very thin extracted sheet of plasmodium intact into the reactivating solution which contained ATP and $MgCl_2$. They observed contraction of certain areas of the extracted plasmodium. In the model plasmodium it was mainly

---

*Renaud et al. (1968) found a close resemblance between the amino acid composition of dynein and actin.

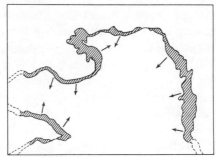

Fig. 21. Contraction in part of a model of myxomycete plasmodium under the influence of ATP (from: Kamiya and Kuroda, 1965). Outer line shows boundary before immersion in solution with ATP, inside line boundary in solution with ATP; arrows show directions of contraction of model.

the thin bands, into which the plasmodia branched, under the experimental conditions used, in the peripheral parts which contracted. These bands became thinner under the influence of ATP and Mg^{++}, i.e., they contracted transversely. The degree of contraction varied in different places in the same preparation, and the maximal linear contraction was about 30% of the initial linear size of this area. The contraction took place rapidly and was complete in 2 min. It was anisodimensional, not only in the peripheral zone consisting of fan-shaped projections or outgrowths, but also in the unbranched part: here also the contraction took place perpendicularly to the outside edge of the plasmodium, and no shortening of the bands took place (Fig. 21).

This picture of the contraction of the glycerinated model supports the hypothesis that the cause of movement of the plasmodial endoplasm is the sum of the successive contractions of its ectoplasm; contraction of the latter is peristaltic in character, as a result of which the endoplasm flows in a certain direction. However, a flow of cytoplasm has not yet been observed in myxomycete models.

Recently, repetitive rhythmic cycles of contraction and relaxation have been obtained on plasmodial models of the myxomycete Physarum polycephalum (Kamiya, 1968). Kamiya and Kuroda excised thin strands from the plasmodium, suspended them for 1 h

vertically in a humid chamber with a load on the free lower end
(20 mg/cm^2), and obtained orientation of the contractile fibrils of
the plasmodium along the longitudinal axis of the strand. Orientation of the fibrils in this manner was verified electron-microscopically. After a short time the hanging strands began a rhythmic alternation of contraction and relaxation (by 10% of its initial
length) and to lengthen again. Glycerinated models obtained
subsequently from these strands with oriented fibrils also contracted and relaxed rhythmically, like the living, stretched
strand, by the action of $(1-2) \times 10^{-3}$ M ATP and $(1-10) \times 10^{-8}$
g-ion/liter Ca^{++}. The contraction–relaxation cycle, just as
for living plasmodia, took about 1.5 min. Probably this rhythmic contraction–relaxation of the strands reflects the oscillatory character of the cytoplasmic currents of the plasmodium
itself, in which the cytoplasm moves at equal time intervals first
in one direction, and then in the other.

As a substance inducing contraction in model plasmodia ATP
cannot be replaced either by AMP or by inorganic phosphate.

After Lehninger (1959a, 1959b) had shown how swollen mitochondria can be made to contract by the action of ATP and Mg^{++},
attempts were made, first, to isolate a contractile protein from
mitochondria and, second, to prepare contractile glycerol-extracted models of mitochondria. Both of these aims have been achieved.
A protein with contractile properties, similar to those of actomyosin and consisting of two (myosin-like and actin-like) proteins was
obtained from mitochondria (Ohnishi and Ohnishi, 1962a, 1962b;
Kazakova and Niefakh, 1963).

However, attention must be drawn to some recent work by
Bemis et al. (1968), the results of which appear to contradict the
view that contractile proteins exist in mitochondria. These workers showed that 0.6 N KCl, under various conditions, does not extract a protein from liver mitochondria which could be superprecipitated from ATP and Mg^{++} or from ATP without Mg^{++}. The
same workers extracted such a protein from the sarcosomes of
the heart. They consider that this protein results from contamination of the sarcosomes of heart muscle by actomyosin, and that
because of this they were unable to obtain it from nonmuscle (liver)
mitochondria. Bemis and co-workers emphasize the absence of
any form of fibrillary material in sarcosomes. Nevertheless,

Nakazawa (1964) obtained glycerinated models of mitochondria which respond to ATP and Mg^{++} by contraction, and which hydrolyze ATP in the process. Mitochondria thus contract by means of a mechanism similar to the contractile mechanism of muscles. Admittedly, the work of Kazakova (1964) showed that glycerinated mitochondria, swell, and do not contract, under the influence of ATP. These discrepancies between the results obtained by Nakazawa and Kazakova are surprising, because they used similar methods to investigate the same objects.

The mechanism of chloroplast contractility has been investigated with the aid of models. Under the influence of light chloroplasts undergo structural changes and contract. Just as with mitochondria, a contractile protein has been extracted from them (Packer and Marchant, 1964), and glycerol-extracted contractile models, contracting in the presence of ATP and Mg^{++} have subsequently been obtained (Packer and Young, 1965). Consequently, the chloroplast, an organoid of the plant cell, also has a contractile system which functions by the same principle as the muscle cell.

Several forms of contraction in nonmuscle cells and their elements (rounding of fibroblasts and sarcoma cells, contraction of amebas, mitochondria, and chloroplasts) thus share a common mechanism of contraction with muscles: enzymic interaction between the contractile protein and ATP. Experiments on cell models and models of organoids have made it possible to analyze and compare the mechanisms of the contractile responses of the objects mentioned.

The examples we have discussed do not exhaust the list of contractile mechanisms which are now known to be based on the same principle as the mechanism of contraction of the muscle fiber, namely interaction between the contractile protein and ATP. Other examples are not mentioned here because they are cases in which movement arises as the result of a combination of several mechanical reactions, only one of which is contraction by the influence of ATP. These will be discussed below (see page 74).

2. LENGTHENING INDUCED BY ATP
(Hoffmann-Berling, 1954a, 1954b)

Models of fibroblasts which, under the conditions described above, contract under the influence of ATP and can be made to lengthen, also as a result of a reaction with ATP, under different conditions. This takes place primarily in the presence of high, superoptimal ATP concentrations, which inhibit hydrolysis and contraction through enzyme-substrate inhibition.

All other factors which inhibit the contraction of a model also facilitate its lengthening (stretching) in response to ATP. Such factors include increased ionic strength of the reactivating solution, urea, which acts as a plasticizer, and a short extraction time, not long enough for destruction of the relaxation system inhibiting contraction of the model to take place. Complete removal of Ca^{++} and Mg^{++} from the medium, although preventing contraction of models, does not inhibit their stretching. Addition of Ca^{++} to the medium, on the other hand, can abolish stretching of a model: Ca^{++} ions inhibit the relaxation system. Contractile poisons, such as Salyrgan and Germanin, do not inhibit lengthening. The presence of protamine sulfate in the medium in a concentration of 1×10^{-5} g/ml, which does not prevent contraction, on the other hand does prevent lengthening, and it thus behaves as a specific inhibitor of lengthening (of relaxation). Some degree of lengthening (admittedly incomplete) can be induced also by inorganic pyrophosphate, which is not broken down in the process. These facts suggest that the relaxation induced by ATP is based on its plasticizing action. Be that as it may, it is absolutely clear that relaxation takes place by a different mechanism from that of contraction. It is based on the binding of ATP without its hydrolysis. It must be assumed that lengthening of the model (and, consequently, of the living cell) is not an active process, for it is not accompanied by the breakdown of ATP or of high-energy compounds. By acting as a plasticizer, ATP probably abolishes the rigidity of the contractile elements of the cytoplasm, so that the cell can once again stretch on account of the elastic forces of the previously contracted structures, i.e., on the basis of entropy.

ATP-induced relaxation can be reproduced in models obtained from cells in the prophase stage, and also in interphase models after their preliminary contraction by ATP. By raising the ATP concentration to 4×10^{-2} M, and also by combining ATP in a physiological concentration with urea (2.5×10^{-3} M ATP + 0.8 M urea) previously contracted models can be made to relax. Models prepared from cells in metaphase, anaphase, and telophase will also lengthen under the same conditions.

Lengthening (stretching) is only one component of the movements taking place at the stages of anaphase and telophase. The other component of the mechanical activity taking place in these two stages of mitosis is contraction. The two processes run a parallel course and are coordinated, and only their combination can lead to a normal biological effect. We shall therefore examine these movements in the following sections, paying special attention to each of the stages of mitosis.

3. MOVEMENTS COMPOSED OF A SINGLE COMBINATION OF ATP-INDUCED CONTRACTION AND RELAXATION

Let it be said at once that this section is extremely conventional and that, strictly speaking, no valid case can be made out for it. Essentially any combination of movements described in it should be subdivided into its component elements and each of them examined separately in the appropriate preceding sections: on contraction under the influence of ATP and on relaxation through binding with ATP. In the case of anaphase, contraction and lengthening are actually carried out by different motile structures. The motor activity of the muscle fiber is, in the last resort, a combination of contraction and relaxation performed by the same structures but under different conditions. However, it is examined in the section on contraction under the influence of ATP.

The present section is considered on its own entirely because the movements described in it are combinations of motor acts taking place simultaneously, but in opposite directions, and it is only when so combined that they serve their biological purpose.

Movements in Anaphase of Mitosis

During anaphase the cell elongates and at the same time homologous sets of chromosomes separate and move toward the opposite poles of the cell. This separation of the chromosomes is brought about by the spindle system, as a result of a combination of two different motile processes. The first of these processes is contraction of the stretched chromosomes. The second is lengthening of the central spindle, i.e., the part of the spindle which connects the two centrosomes. As a result of the first of these two movements, i.e., contraction, the chromosomes are brought closer to the corresponding centrosome as if drawn to it. As a result of

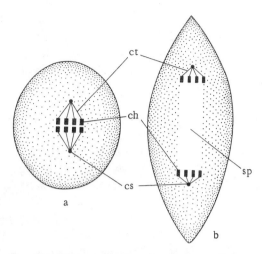

Fig. 22. Diagram showing changes in a cell (or in a model) during anaphase (after Hoffmann-Berling, 1954b): a) before beginning of anaphase, b) at completion of anaphase; ch) chromosomes, cs) centrosomes, ct) chromosome threads of spindle, sp) central spindle.

the second movement, i.e., lengthening of the spindle, the centrosomes themselves move away from each other toward the poles of the cell, taking the attached chromosomes with them.

Usually the chromosomes separate through the successive performance of both these movements of the spindle system against the background of elongation of the cell body (Fig. 22). However, their separate role in the final result, that of separation of the chromosomes to the opposite ends of the elongated cell, differs from one cell to another. The extreme cases are those in which separation of the chromosomes takes place entirely through contraction of the chromosome threads of the spindle (staminal hairs of Tradescantia) or, conversely, entirely by lengthening of the spindle itself (spermatocytes I of the aphis Tamilia). In the spermatocytes of the Orthoptera the two movements take place simultaneously, while in spermatocytes of the Hemiptera and Homoptera, they take place consecutively.

All three types of movement characteristic of anaphase cells (elongation of the cell body, contraction of the chromosome threads of the spindle, and lengthening of the central spindle) are reproduced in their models. In models of fibroblasts obtained by glycerol extraction in early anaphase and then placed in medium containing ATP and Mg^{++}, the sets of chromosomes separate. This takes place under conditions which ordinarily induce contraction of all models. It can accordingly be postulated that the separation of chromosomes toward the centrosomes in such models takes place through contraction of the chromosome threads (Hoffmann-Berling, 1954b). The treads themselves cannot be seen in unfixed models, and it is clear that experiments should be carried out with appropriate staining of anaphase models fixed before and after addition of ATP. This would provide direct evidence in support of contraction of the threads in early anaphase under the influence of ATP. It is unfortunate that Mazia's (1961) preparations of the isolated mitotic apparatus, including those made with dithioglycol, cannot yet be obtained free from functional impairment, and they cannot therefore be used to study the physiology of mitotic movements.*

   It is difficult to obtain elongation of the cell body concurrently with contraction of the chromosome threads in the same model (as occurs in living cells), for elongation of the body requires conditions different from those usually obtaining in a model during lengthening. Even where success is obtained, it is only with briefly extracted models in which the relaxation system still remains intact. In living anaphase cells it is evidently unevenly distributed; it is absent in the region of the chromosome threads and does not therefore prevent them fron contracting.

   The mechanism of contraction of the chromosome threads is probably related to the mechanism of contraction of interphase cells and muscles. Evidence in support of this kinship is given by the fact that the birefringence of the chromosome threads of the spindle decreases during movement of the chromosomes toward the centrosomes, like the birefringence of myofibrils during their contraction.

   In models obtained from fibroblasts in the stage of middle anaphase, lengthening of the spindle (resulting in separation of the centrosomes and, passively, of the chromosomes) concurrently with elongation of the bodies of the models can be obtained by the action of ATP (Figs. 23 and 24). However, simulation of lengthening of the central spindle (the "propulsion body" of Belar) re-

---
*See Notes Added in Proof, page 170.

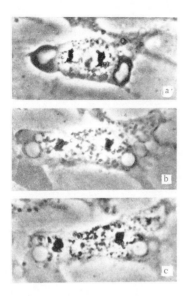

Fig. 23. Anaphase models of subcutaneous chick fibroblasts (from Hoffmann-Berling, 1954b): Briefly (for 1-2 h) extracted models (a) without ATP, (b) 3 min after immersion in solution containing 2.5×10^{-3} M ATP and 0.8 M urea, and (c) after 10 min in the same solution; $\mu = 0.20$, pH 7.2.

quires different conditions from contraction of chromosome threads. Before it can be lengthened the structure must be plasticized, either by application of ATP in high concentractions ($2-4 \times 10^{-2}$ M) or by addition of urea to the medium (0.8 M urea + $1-2 \times 10^{-3}$ M ATP), or by increasing the ionic strength of the medium to $\mu = 0.4$ (Figs. 25 and 26). These illustrations show that the addition of urea as a plasticizer to the medium can reduce the ATP concentration and the ionic strength of the solution to physiological values, and in this way lengthening of the spindle can take place. It is evident that in the living cell the role of plasticizer is played not by urea, but by the physiological relaxing factor.

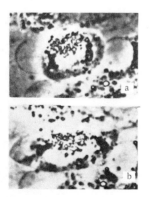

Fig. 24. Anaphase models of subcutaneous chick fibroblasts (from Hoffmann-Berling, 1954b). Models from cells in middle stage of anaphase (a) without ATP and (b) 5 min after immersion in solution containing 1.5×10^{-3} M ATP; $\mu = 0.21$, pH 7.0, t = 37°C. Lengthening of body of model and separation of sets of chromosomes toward opposite poles can be seen in Figs. 23b, 23c, and 24b.

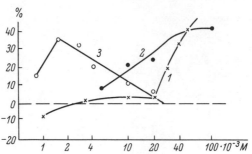

Fig. 25. Separation of sets of chromosomes in anaphase models (subcutaneous fibroblasts of a chicken) as a function of ATP concentration and of plasticization by urea (from Hoffmann-Berling, 1954b). Abscissa, ATP concentration; ordinate, change in distance between sets of chromosomes (in % of initial distance). 1) Separation under the influence of ATP without urea; 2) after 5 min at 0°C in 0.8 M urea without ATP, and then in ATP without urea; 3) preliminary treatment with 0.8 M urea, then with ATP together with urea. ATP as a plasticizing agent can be partially replaced by another plasticizing agent, urea.

In model cells the lengthening of the spindle may be so great that the chromosomes are expelled outside the cell (Fig. 27). If the remnants of cytoplasm in the model exert strong resistance to separation of the chromosomes, the central spindle is twisted during movement of the model. If this resistance occurs only on one side, the central spindle can be bent, so that the chromosomes deviate the side (Fig. 28). Anaphase lengthening movements in

Fig. 26. Effect of ionic strength and of urea (i.e., plasticization) on separation of sets of chromosomes in anaphase models of subcutaneous fibroblasts (from Hoffmann-Berling, 1954b). Abscissa, ionic strength (in relative units), ordinate, change in distance between sets of chromosomes (in % of the initial level). 1) Without urea, plasticization obtained only at a high ionic strength; 2) plasticization with 0.8 M urea. Increased ionic strength facilitates separation of chromosomes by plasticizing the model; the plasticizing action of urea. Reactivation with 2.5×10^{-3} M ATP.

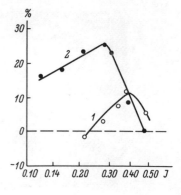

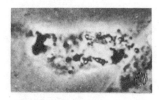

Fig. 27. Anaphase model of subcutaneous chick fibroblasts (from Hoffmann-Berling, 1954b) 7 min after immersion in solution containing 5×10^{-3} M ATP, t = 37°C; lengthening of central spindle proceeding out of control, with no restraint, so that chromosomes leave the cell model.

models (lengthening of the body of the model and also of the spindle) are most marked in the temperature range 25-37°C (Fig. 29).

To explain the coordination between the mechanisms of separation of the chromosomes in anaphase and also of lengthening of the cell itself, it is assumed that there is an uneven distribution in the cell structures of the factor inhibiting contractility, i.e., of the relaxing factor, which regulates the distribution of free calcium ions in the cell. As a result the different elements of the cell simultaneously perform opposite movements. In model experiments as a rule it is impossible to obtain contraction of the chromosome threads of the spindle without simultaneous contraction of the cell itself, with the result that the spindle is deformed and the picture of the first stage of anaphase is distorted. In a recent investigation Kinoshita and Yazaki (1967) studied the redistribution of the relaxing system in the cell during mitotic division in the sea urchin egg during cleavage. They found that the granular elements of the relaxing system are concentrated around the centrioles during anaphase. Dougherty (1965; cited by Kinoshita and Yazaki, 1967) found that the smooth endoplasmic

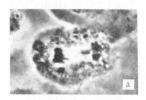
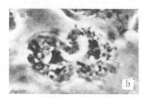

Fig. 28. Anaphase models of subcutaneous chick fibroblasts (from Hoffmann-Berling, 1954b). Model of cell in late stage of anaphase (a) without ATP and (b) 6 min after immersion in solution containing 2.5×10^{-3} M ATP, t = 22°C; central spindle, unable to overcome resistance of coagulated cytoplasm, has curled up and chromosomes have moved sideways; commencing cytokinesis can be seen. Sets of chromosomes are blackened in all photographs in Figs. 23, 24, 27 and 28.

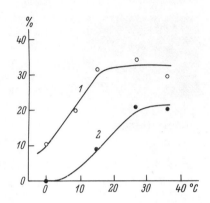

Fig. 29. Lengthening of anaphase cell model (1) and model of separation of chromosomes (2) as a function of temperature (from Hoffmann-Berling, 1954b). Models of subcutaneous fibroblasts of a chicken. Effect recorded 4 min after addition of 2.5×10^{-3} M ATP. Abscissa, temperature (in °C); ordinate, increase in length of cell model or increase in distance between sets of chromosomes (in % of initial values).

reticulum in hepatocytes of rodents, which can exert a relaxing effect on the contractile elements of the cell, migrates during metaphase and anaphase toward the poles of the cell.

Since it is the elements of the smooth endoplasmic reticulum which, by actively absorbing calcium ions, play the role of relaxing system in the cell, migration of the reticulum toward the poles can be regarded as the condition directly determining lengthening of the polar regions of the cell. The results of model experiments suggest that the chromosomes themselves play only a passive role in movement. When, in experiments on models, they slipped off the spindle, the latter continued to lengthen and the chromosomes remained in the same place.

The mechanisms of movements taking place during anaphase have thus been analyzed with the aid of glycerinated cell models. Model experiments have shed light on the phenomenon of contraction of the chromosome threads of the spindle, the mechanism of which is similar to that of muscle contraction and with contraction of interphase or prophase cells; it takes place as the result of hydrolysis of ATP. At the same time, they have demonstrated the existence of a movement in the opposite direction, also with a role in anaphase and consisting of relaxation and lengthening of cytoplasmic structures. These movements are related to relaxation of muscles and also to lengthening of previously contracted interphase cells; they are due to binding of ATP without its hydrolysis. Anaphase models can also be used to analyze the mechanisms of coordination on the movements takin place during this phase of mitosis.

Movements in Telophase of Mitosis

(Hoffmann-Berling, 1954c, 1964; Kinoshita and Hoffmann-Berling, 1964)

Division of the cell into two daughter cells takes place through the contractile formation of a central constriction of the cytoplasm (cytokinesis), but contraction is not the only movement taking place during telophase. The cell simultaneously increases in length (Fig. 30). Both processes are reproduced in glycerinated models made from cells in the late stage of division of the nucleus (Figs. 31 and 32). As a matter of fact, cytokinesis never goes to completion in models: a narrow band of cytoplasm joining the daughter cells always remains. Admittedly, the appearance of this zone is altered in the presence of ATP when observed under the phase-contrast microscope: a reticular structure appears in the cytoplasm, which initially was optically empty, but what this means is at present unknown.

The conditions inducing the formation of a central contractile construction band in the model are identical with those inducing rounding of interphase models and contraction of muscle models. Cytokinesis of the model can also be induced by ATP in concentrations coinciding with those normally found in the cell or muscle fiber. Magnesium ions are also necessary for this to take place, and just as in other types of model contraction, it must be present in very small concentrations. Cytokinesis is also prevented by inhibitors of cell and muscle contraction, containing heavy metals:

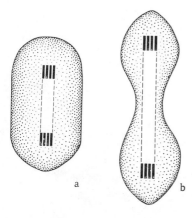

Fig. 30. Diagram illustrating changes in the cell (or model) during telophase (after Hoffmann-Berling, 1954c): a) before beginning of telophase, b) during telophase.

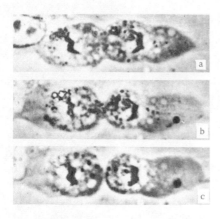

Fig. 31. Cytokinesis of telophase models under the influence of ATP (from Hoffmann-Berling, 1954c). Model of chick subcutaneous fibroblast (a) without ATP, (b) 5 min after immersion in solution containing 2.5×10^{-3} M ATP, t = 20°C, and (c) after 20 min in the same solution.

2×10^{-4} M Salyrgan and 5×10^{-4} M Oxarsan, the action of which can subsequently be abolished by 2×10^{-2} M cysteine; weak thiol poisons, such as 1×10^{-2} M monoiodoacetate, likewise do not act on this type of model contraction. Contraction of the equator of telophase models depends on the temperature, the ionic strength, and pH of the medium in the same way as contraction of interphase cell models and muscle models.

Cytokinesis and lengthening of the telophase model of the fibroblast are very different functions of ATP concentration, so that each one of these movements can be obtained separately from the other. In high concentrations of ATP (with ordinary values of pH and μ, [ATP] $> 1 \times 10^{-2}$ M) lengthening of the body of the model is activated, but contraction and the formation of an equatorial constriction are arrested (Fig. 33). Under these conditions, if a

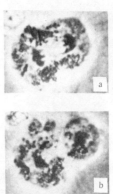

Fig. 32. Cytokinesis of tlophase models under the influence of ATP (from Hoffmann-Berling, 1954c). Model of chick scleroblast (a) without ATP and (b) 12 min after immersion in solution containing 1.5×10^{-3} M ATP; $\mu = 0.12$, pH 7.1, t = 37°C. Sets of chromosomes are blackened in all photographs in Figs. 31 and 32.

central constriction has started to form, not only does the process stop, but it is replaced by widening of the equatorial zone. This widening is not abolished by 2×10^{-4} M Salyrgan. It can also be induced by inorganic pyrophosphate in concentrations exceeding 1×10^{-2}; it is promoted by plasticizers such as 0.8 M urea. The phenomenon of widening of the equatorial zone is probably based on the relaxation mechanism depending on the plasticizing action of ATP, i.e., on the binding of ATP without its hydrolysis — the same mechanism as was discussed when we examined relaxation without its hydrolysis — the same mechanism as was discussed when we examined relaxation (lengthening) of many other models.

Conversely, in reduced ATP concentrations [$(2-5) \times 10^{-3}$ M] contraction takes place, but not only in the form of an equatorial constriction band; as curve 2 in Fig. 34 shows, contraction also takes place at the poles, i.e., the length of the cell is reduced. This latter effect becomes predominant, the model cell is mechanically incapable of equatorial constriction, and it simply becomes rounded and does not divide, thus behaving like an interphase model.

In a physiological concentration, ATP causes contraction in the center and lengthening at its poles, leading to an increase in the length of the model cell. In other words, under these conditions the model behaves just like a living cell during telophase.

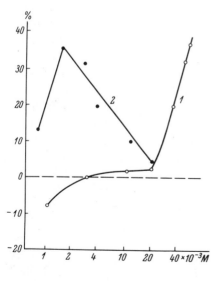

Fig. 33. Change in size of telophase model as a function of ATP concentration (from Hoffmann-Berling, 1954c). Abscissa, ATP concentration; ordinate, change in size of model (in % of initial): 1) width at equator, 2) length of model. Models of subcutaneous fibroblasts, reactivation for 7 min at 37°C.

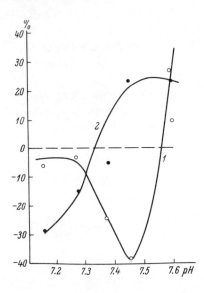

Fig. 34. Changes in size of telophase model as a function of pH during reactivation of model by 2.5×10^{-3} M ATP (from Hoffmann-Berling, 1954c). Abscissa, pH; ordinate, relative changes in size of model (in % of initial): 1) width at equator, 2) length of model. Models of subcutaneous fibroblasts, reactivation for 7 min at 37°C.

An increase in the concentration of ATP causes inhibition of contractility of all parts of the model, but its pole is more sensitive to this action of ATP than the equatorial regions. This difference enables the model cell to contract at the equator and to lengthen at the poles simultaneously. ATP in physiological concentration causes contraction of the equatorial part of the model (in the living cell), but is superoptimal for the contractility of the poles, so that relaxation of the polar regions takes place and, as a result, the whole model is lengthened (Fig. 33). To judge from the reaction of contractility of the equator and the increase in length of the model itself to a change in pH of the medium (Fig. 34), the sensitivity of different parts of the model varies with respect to this factor also. At low pH values, ATP causes contraction in all parts, the model contracts in all directions, and cytokinesis cannot take place. Average pH values are optimal at the same time for contraction in the equatorial region and for relaxation at the poles, i.e., for lengthening of the body of the model; cytokinesis takes place. At high pH values, only relaxation develops throughout the model under the influence of ATP: the model lengthens, but in the equatorial region it widens; cytokinesis does not take place.

The unequal sensitivity of different parts of the cell of ATP, leading to the appearance of contraction and lengthening, can be attributed to the irregularity of distribution of relaxing factor throughout the cell. During telophase it is active at the poles, but absent in the equatorial region or, perhaps, it is inactive.

As a result, even if the ATP distribution in the cell is uniform and the contractile system is present in all parts of the cell, the poles remain relaxed while the contractile system of the equatorial region contracts. An equatorial constriction is formed in the cytoplasm. The relaxing system is redistributed evidently while anaphase takes place, since cytokinesis can be obtained also in models of cells which are in middle anaphase.

In models of fibroblasts which are extracted for more than 48 h the relaxing system is extracted or injured. Such models accordingly react to ATP equally at the poles and equator of the cell, both of which contract because there is nothing to prevent it. The model becomes regularly round like an interphase model. The same events take place after brief extraction, i.e., in models preserving the endogenous relaxing factor, if they are placed in medium containing an excess of Ca^{++}. Under these circumstances the action of relaxing factor in binding Ca^{++} is blocked or, more exactly, the relaxing system is saturated to the limit with calcium and cannot bind all the calcium present. As a result the concentration of free calcium near the motile structures of the model exceeds 1×10^{-6} g-ions/liter, and the mechanochemical reaction of the contractile system with ATP is activated.

Addition of "relaxing granules" obtained from muscles or fibroblasts to model fibroblasts in which relaxing factor is absent as a result of prolonged extraction inhibits contractile activity once again at the cell poles. Local contraction to form an equatorial constriction of the cytoplasm takes place in models in a medium containing ATP and relaxing granules, and under optimal conditions the model cell divides into two daughter cells which are joined together only by a narrow bridge of cytoplasm (Kinoshita and Hoffmann-Berling, 1964).

Not only Ca^{++}, but any agent which prevents the functional activity of the added relaxing granules, will induce general contraction of a cell model, making equatorial constriction of the cytoplasm impossible.

Hoffmann-Berling (1964) claims that the models still retain, although in an inactivated state, their own endogenous relaxing system, which can be reactivated by the addition of exogenous relaxing granules. He considers that these exogenous granules liberate a soluble cofactor which reactivates the endogenous relaxing system.

Any agent which inactivates the endogenous relaxing system will induce a general contraction of the cell model, and in this case, just as when exogenous granules are inactivated, no equatorial constriction of the cytoplasm can take place. This is what happens in experiments in which cell models were treated with Salyrgan together with cysteine (this treatment inactivates the endogenous relaxing system). It also happens when calcium salts are added to the medium, for calcium ions inactivate both the endogenous system and the exogenous relaxing granules (Kinoshita and Hoffmann-Berling, 1964).

If only a small quantity of granules is added, the process of contraction takes place also at the poles of the cell model; in that case a general contraction of the cell without an equatorial constriction of the cytoplasm develops. Cytokinesis of the model does not take place likewise if granules are added to excess: under these circumstances they have a relaxing action also on the equatorial region, as a result of which no contraction takes place there. Prevention of cytokinesis by the action of calcium ions is also regarded as the result of inactivation both of the endogenous relaxing system and of the added relaxing granules.

It could be concluded from the experiments of Hoffmann-Berling and Kinoshita to study the action of relaxing granules on model fibroblasts that, although the mechanism of action of granules is the same in muscles and fibroblasts, in the latter the contraction is local, unlike in muscles. Local contraction of the equatorial region of the cell during telophase takes place because the activity of the relaxing system is distributed nonuniformly between the polar and equatorial regions. However, until recently there was no general agreement on the cause of the local contraction of the equatorial region: whether it was a decrease in the sensitivity of its contractile elements to the action of the relaxing system or whether this system itself was less highly developed in the equatorial region. For this reason, Kinoshita and Yazaki (1967), as has already been mentioned, investigated the localization of the intracellular relaxing system during the mitotic cycle, specifically during cleavage of the sea urchin egg. These workers obtained a model of an egg during cleavage and prepared an antiserum against relaxing granules obtained from this object, i.e., against the relaxing system of the sea urchin egg. They labeled the antibody with fluorochrome and then injected it into model eggs at different stages of cleavage. They then determined the localization of the fluorescent antibody, which was the same as the localization of the relaxing system. In this way Kinoshita and Yazaki demonstrated and studied migration of the cellular relaxing system during cleavage of the egg. They found that the relaxing system in the eggs of Clypeaster japonicus is attracted to the mitotic apparatus and concentrated in anaphase in the region of the centrioles, but later it disappears completely in the region of the future furrow, but meanwhile it develops intensively in the polar regions. These findings, according to Kinoshita and Yazaki, suggest that disappearance of the relaxing system in the equatorial region is the cause of contraction of this region, i.e., of the appearance of a cleavage furrow. However, they also describe the results of similar experiments with models of the cleaving eggs of another species of sea urchin Pseudocentrotus depressus, and they point out that in this case the pattern of migration differs significantly from that described above. During anaphase the relaxing system appears at the poles, but later, during telophase, it is found in the equatorial region, for example, where the cleavage furrow is formed. Admittedly, these workers emphasize that the pattern of redistribution of the relaxing system in eggs of Pseudocentrotus, by contrast with the eggs of Clypeaster, was indistinct, and when describing it they are merely stating a general tendency; on the other hand, they were never able to observe the opposite tendency.

The most significant fact confirming the hypothesis that no relaxing system exists in the region of formation of the equatorial constriction in the cytoplasm as the factor permitting contraction to take place in that area was thus discovered in only one of the two objects tested. The only difference between the conditions of the experiments on the eggs of these two species of sea urchin was that homologous serum was injected into the model eggs of Pseudocentrotus, while heterologous serum was injected into the Clypeaster eggs, since relaxing granules from Pseudocentrotus eggs had been used as the antigen to prepare it. Kinoshita and Yazaki also cite the observations of Dougherty (1965; cited by Kinoshita and Yazaki, 1967) to the effect that the smooth endoplasmatic reticulum in the hepatocytes of rodents migrates toward the poles during metaphase and anaphase.

Kinoshita (1968) induced artificial parthenogenesis in sea urchin eggs by means of a hypertonic solution after preliminary centrifugation. Cleavage and cytokinesis of the eggs were not preceded by the formation of a mitotic apparatus. These experiments also showed that the cleavage furrow appears in areas of the egg without any endoplasmic reticulum: disappearance of elements of the reticulum actively removing Ca^{++} from the surrounding cytoplasm created conditions suitable for local contraction in the form of an equatorial constriction of the egg.

Local control over the ability of the cleaving egg to contract locally may be exerted by the endoplasmic reticulum through its liberation of a heparin-like sulfo acid into the hyaloplasm, after which the acid is reimbibed into its vesicles and bound again. Kinoshita (1969) found that heparin in sea urchin eggs is partly bound with the endoplasmic reticulum, and partly in a free state; during cleavage of the egg the quantity of heparin in the two forms fluctuates reciprocally. An increase in the content of free heparin is accompanied by a decrease in density of the cytoplasm, i.e., by its plasticization; binding of heparin by the relaxing system, on the other hand, is accompanied by an increase in the rigidity of the cytoplasm.

It seems both desirable and possible to perform an experiment similar to that of Ridgway and Ashley (see page 32), not on a muscle fiber stimulated electrically, but on a dividing cell in the stage of anaphase or telophase. Such an experiment could consist of the injection of a protein which fluoresces in the presence of $1 \times 10^{-6} - 1 \times 10^{-5}$ g-ion/liter Ca^{++} into, for example, a cleaving sea urchin egg and recording the intensity and distribution of fluorescence in the egg during the development of mitosis or even during anaphase or telophase only. If such an experiment were successful, it would show conclusively whether in fact the appearance of Ca^{++} precedes local contraction and its disappearance precedes lengthening.

Models have thus revealed the mechanisms of movements taking place during telophase and leading to cytokinesis and have enabled these mechanisms to be analyzed. Experiments with models have shown that there are two separate processes: contraction of the cortical layer of the equatorial region, which is related to contraction of muscle and of other cells, during which hydrolysis of ATP takes place, and a process of relaxation at the poles of the cell, related to that leading to relaxation of muscle, lengthening of cell bodies in anaphase, and lengthening of their central spindle and which is due to binding of ATP without its hydrolysis. Model experiments have demonstrated the mechanism of two types of movement, taking place simultaneously but in opposite directions: contraction and relaxation. Models, however, have not yet provided the key to the elucidation of the mechanism of the last stage of cytokinesis: rupture of the thin cytoplasmic band joining the newly formed daughter cells.

In all cases described above, however much they differ from each other, there is an important common feature: movement is induced by ATP, although the character of its interaction with the protein structure and, correspondingly, the character of the movement itself differ from one case to another. However, in protozoans cytoplasmic movements in which either active contraction or active lengthening takes place without the participation of ATP have been discovered and, to some extent, studied.

4. CONTRACTION INDUCED BY CALCIUM IONS AND LENGTHENING INDUCED BY ATP

The stalk of <u>Vorticella</u> has some properties in common with muscle fibers. In the myoneme (spasmoneme), the contractile element of the stalk, there are submicroscopic fibrils; the myoneme has positive birefringence, the magnitude of which, just as in the muscle fiber, decreases during contraction, and the contraction of the stalk itself is easily evoked by electrical stimulation; the rate of its shortening is of the same order as that of cross-striated muscle.

Levine (1956) extracted vorticellas with glycerol and obtained models of the stalks. Three species were chosen: Vorticella campanula, V. nebulifera, and V. convallaria. The technique of preparation of the models had one distinctive feature: Ca^{++} was removed from the extracting solution, and even the animals were washed with versene (EDTA) before they were immersed in the extracting solution. The experiments showed that the myoneme of the model becomes coiled (and as a result, the stalk contracts) on the addition of Ca^{++} to the medium. To evoke contraction all that was necessary was to place the model stalks in a medium containing as little as 5×10^{-4} M $CaCl_2$ (together with 0.5 N KCl). As soon as the stalks, when contracted by the action of Ca^{++}, were transferred to a solution containing 4×10^{-3} M versene, which binds Ca^{++}, the stalks straightened again. By transferring the model from versene solution into calcium solution and vice versa, complete cycles of contraction and relaxation could be obtained many times over. Levine observed contraction of model stalks as a result of the action of other bivalent ions (Mg^{++} and Mn^{++}), but they were less effective than Ca^{++}. In Levine's experiments ATP, in conjunction with Mg^{++}, caused neither contraction nor relaxation of the stalk. He concluded that Ca^{++} activates the factor responsible for coiling of the myoneme both in the model and also, evidently, in the living cell.

Hoffmann-Berling (1958) obtained models of the stalk of Vorticella gracilis by using a different extraction technique. Like Levine, he removed all Ca^{++} from the extracting solution with versene, but in addition, he also excluded glycerol from it and instead, he added saponin. Saponin was added to produce lysis of the stalk membrane and in this way to facilitate extraction of substances soluble in 0.12 N KCl, and to provide unobstructed entry of the test substances (Ca^{++}, ATP, etc.) into the stalk.

Hoffmann-Berling kept his cells in the saponin solution for only 20-30 min before the model was ready to use. He then carried out his experiment either at once, having washed all the saponin from the model, or transferred it to preserving solution containing glycerol as the preservative.

The Vorticella models thus obtained contracted instantaneously in medium containing Ca^{++} (Fig. 35), Sr^{++}, or certain ammonium compounds, such as cetyl-trimethylammonium chloride.

Ca^{++} and Sr^{++} produce maximal shortening of the stalk in concentrations below 1×10^{-5} M. The use of calcium or strontium salts in higher concentrations increases the rate of contraction but not its amplitude. During contraction tension develops, and this may be so great that if the stalks are left in calcium solution, i.e., in a state of contraction, the myoneme, the contractile element of the stalk, breaks up into parts. The models relax if Ca^{++} is removed somehow or other from the solution. This result is best achieved with the aid of versene and hexametaphosphate. These substances, in concentrations as low as $10^{-4}-10^{-3}$ M, produce total relaxation of the stalk. By transferring the models from solution containing versene into solution containing calcium and vice versa, Hoffmann-Berling, like Levine, obtained several (as many as 12) complete cycles of contraction and relaxation, with no diminution in the degree or speed of shortening. Sulfhydril poisons, Salyrgan, and Germanin, which poison muscle and fibroblast models, have no effect either on the calcium-induced contraction or the versene-induced relaxation of the stalks. These poisons, incidentally, were tested in concentrations 10 times higher than usual. The results of experiments with sulfhydryl poisons provide some of the evidence that the cause of contraction of the stalk model in contact with Ca^{++} is not residual traces of endogenous ATP, traces which could be suspected of being present in the model. In that case the poisons would have suppressed contractility, and this did not take place. Another indication that traces of ATP are not concerned in the contraction is that repeated contractions of the same stalk, as mentioned above, were not weaker than the first contraction. Con-

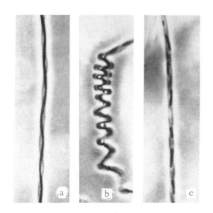

Fig. 35. Model of stalk of <u>Vorticella</u> (from Hoffmann-Berling, 1958) (a) in solution with versene (EDTA) and (b) after removal of versene and addition of 5×10^{-5} g-ion/liter Ca^{++}. To prevent rupture of the stalk, ionic strength has been increased ($\mu = 0.3$) and this limits development of tension: (c) Rupture of internal structures of the stalk after free, unrestricted tension produced by Ca^{++} in solution of low ionic strength ($\mu = 0.10$).

sequently, Ca^{++} ions cause contraction of the stalk by reacting with its contractile system. Besides Ca^{++}, the same effect was obtained in Hoffmann-Berling's experiments by the use of Sr^{++}, Ba^{++} ions had a weak action, and Be^{++} and Mg^{++} no action whatever. So far as Mg^{++} is concerned, this result does not agree with Levine's findings. Contraction was also evoked by organic compounds of the cationic detergent type, in which the positive ion has a long nonpolar "tail," and the longer this residue, the lower the concentration of the compound which produces maximal contraction. Contraction induced by cationic detergents differs from calcium contraction: it is slower and its amplitude is lower, and the tension is weaker (the myoneme does not break), and the contraction is irreversible. It is, of course, only of theoretical interest, not being physiological. The physiological agent inducing contraction is evidently calcium ions. The following facts support this view. It is calcium ions, and not other bivalent ions, which are always present in the cells. They give rise to contraction of the model stalk in lower concentrations than other bivalent ions. In the presence of calcium ions a model stalk develops its maximal strength and rate of contraction. Contraction of the stalk under the influence of calcium ions is reversible.

Hoffmann-Berling gives an electrostatic interpretation of the mechanism of contraction of the myoneme, and likens it to the contraction of Katchalsky's polyacrylic models. Protein structures of the relaxed myoneme, according to Hoffmann-Berling, carry an excess of negative charges, repelling each other and therefore holding the myoneme in a lengthened state. Positive ions neutralize some of these charges, and the structure can then change into a contracted (equilibrium) state. Alternatively, contraction takes place through electrostatic attraction between the negative charges carried by the structure itself and the positive charges of added cations. The fact that not every cation will evoke contraction is attributed by Hoffmann-Berling to the specificity of linking of the cations. The cationic detergents are known to form stable compounds readily with the acid groups of proteins.

The question arises: What causes the physiological lengthening of the stalk? Versene is purely and simply a laboratory agent, and it thus still remained to be explained what substance normally performs its function in the <u>Vorticella</u> stalk.

Hoffmann-Berling's search for the physiological relaxing factor showed that under natural conditions the Vorticella stalk is made to lengthen by ATP. The action of ATP differs in principle from the action of versene. Versene binds all Ca ions and lengthening then takes place — that is all there is to it. To produce a fresh contraction all versene must be washed out of the model and it must again be placed in calcium solution. ATP, when added to the solution containing a stalk which has contracted through the action of Ca^{++} causes it to lengthen, but this is at once followed by a fresh contraction because the Ca^{++} ions still remained in the solution. This contraction is again followed by relaxation because the ATP has not all been used up, and so on. Spontaneous rhythmic contraction and relaxation of the model stalk thus takes place, without any rinsing, in the same solution. The stalk contracts instantaneously but relaxes more slowly, and the frequency reaches 20 cycles per minute; sometimes, just as in the living Vorticella, periods of immobility arise. ATP acts only in the presence of Mg^{++}. Optimal conditions for rhythmic contraction of the stalk are: 1×10^{-4} g-ion/liter Ca^{++}, 2×10^{-3} M ATP, and 4×10^{-3} g-ion/liter Mg^{++}.

The rhythmic and repetitive character of the process suggests that the action of ATP is not simply to bind Ca^{++}, for in that case no second contraction would take place. This conclusion is also supported by other evidence: 1) the affinity of ATP for Ca^{++} is less than that of citrate and oxalate, but nevertheless these substances do not cause relaxation; 2) ATP cannot be replaced by GTP, although this has the same affinity for Ca^{++}; 3) the relaxing action of ATP increases with an increase in the Mg^{++} concentration (up to 5×10^{-3} M $MgCl_2$); since Mg^{++} is bound by ATP more effectively than Ca^{++}, the excess of Mg^{++} would liberate the whole of the Ca^{++} from the ATP, and with an increase in Mg^{++} concentration the calcium contraction would increase, and not the ATP-induced relaxation. It is therefore quite clear that physiological lengthening takes place as the result of the reaction between ATP and the contractile protein and not as the result of the binding of calcium by ATP.

During interaction of the contractile protein with ATP the latter is split to provide energy for the cycle of rhythmic contraction and relaxation of the stalk. However, since the contractile protein has not been isolated from the stalk, the splitting of ATP has not been directly proved. There is, however, indirect but weighty evidence that it actually takes place. The relaxing action of ATP is prevented by Salyrgan, and in concentrations ($5 \times 10^{-3} M$) which inhibit the work of muscle and connective-tissue models, and their activity is unquestionably based on ATP hydrolysis.

The chemical energy of ATP in Vorticella is thus utilized in the relaxation phase. This is the mechanism of muscle contraction inverted, or turned inside out. It must be remembered, in this connection, that the contraction mechanism described above, based on neutralization of excess charges of the contractile structure, was put forward as an explanation of muscle contraction. However, this scheme assumed the presence, not of negative, but of positive charges on the contractile structures, and neutralization of these by the powerful ATP anions gives rise to contraction of the muscle.*

Whatever the true explanations of contraction of the muscle and myoneme, it is evident that they both share a common principle. On the other hand, the example of the myoneme shows that the ATPase mechanism is not the only mechanism of contraction. The other motile mechanism, which has just been described, was discovered by the use of a model technique.

Seravin and co-workers (1965) showed that the Vorticella myoneme is not the only case of calcium-induced contraction. In the absence of ATP and in the presence of $(1-5) \times 10^{-3}$ g-ion/liter Ca^{++} the bodies of models of another infusorian — Spirostomum ambiguum — contract vigorously. Contraction of these models is also induced by Mg^{++}, Ba^{++}, Co^{++}, Cd^{++}, Fe^{+++}, and Al^{+++}. In addition, it can be induced by NAD (7 mg/ml), but not by $NAD \cdot H_2$. However, neither the binding of Ca^{++} with EDTA, nor ATP, nor relaxation factor isolated from frog muscles or from Spirostomum itself, nor the action of 6% urea solution, which plasticizes the muscle model, can induce relaxation of models of this infusorian after calcium contraction. Seravin and co-workers showed that treatment of the models with 0.6 M KCl solution, which extracts myosin from muscles, does not prevent the Spirostomum models from contracting under the influence of Ca^{++}.

---

*See Notes Added in Proof, page 170.

According to the same workers the bodies of glycerol-extracted models of Paramecium caudatum will also contract by 10-15% of their initial length by the action of 0.93×10^{-3} $M\,CaCl_2$. Neither ATP + Mg^{++} nor Mg^{++} or Ba^{++} used separately, will give this effect.

Models of the Vorticella stalk suggested by Hoffmann-Berling have also made an important contribution by widening the range of objects from which motile models can be obtained. It was in experiments on the Vorticella stalk that destruction of the membranes by detergents was first used as a method of preparing models. This method has subsequently found wide application.

5. LENGTHENING INDUCED BY CALCIUM IONS
AND INHIBITED BY ATP

(Hoffmann-Berling, 1960)

When the trichocysts of paramecium are liberated they increase by 6-7 times in length and in volume. The motile part of the trichocyst, that is its stalk, consists of fibrils and possesses birefringence; electron-microscopic studies have shown that it also possesses cross-striction (Beyersdorfer, 1951; Krüger et al., 1952). The liberation of trichocysts by Paramecium can be stopped by means of agents which bind Ca^{++}, such as versene. This demonstrates the important role of Ca^{++} in their lengthening. Paramecia, treated with versene, can be destroyed and the trichocysts concentrated by differential centrifugation. Addition of Ca^{++}, Sr^{++}, or cationic detergents to the isolated trichocysts causes their immediate lengthening. The degree of lengthening is the same as with trichocysts of the living cell. Magnesium ions do not give this effect. ATP prevents the lengthening of trichocysts induced by calcium ions, even though sufficient ions to provoke lengthening of trichocysts in the absence of ATP remain in the solution in a free form. ATP cannot make a trichocyst contract if it is already lengthened. Isolated trichocysts possess ATPase activity (0.3 mM phosphorus/mg protein/min at 22°C), and $(5-10) \times 10^{-4}$ M Salyrgan depresses this activity by 70%. In this case, just as with the Vorticella stalk, Ca^{++} and ATP have opposite effects but the action of these substances differs on the stalk and on the trichocysts. Calcium ions induce contraction of the Vorticella stalk, but lengthening of the paramecium trichocyst. ATP also induces a motor response of lengthening in the stalk, but it can only inhibit the motor response of the trichocyst induced by Ca^{++}, and ATP itself has no motor effect.

6. ATP AND CYTOPLASMIC STREAMS IN PLANT CELLS

The cytoplasm of most plant cells is in continuous movement. Streams of cytoplasm can be seen under the microscope because organoids such as spherosomes, mitochondria, and sometimes chloroplasts and nuclei, become involved in this movement. It is difficult to see anything in common between streams of cytoplasm and muscular contraction. They have no resemblance at all outwardly, but investigations have shown that they share a common principle of energetics. It was from this aspect that Takata (1961) investigated the movement of cytoplasm in the stalk of the unicellular alga Acetabularia calyculus.

In young individuals, before they have acquired a cap, the cytoplasm in the stalk moves energetically. Under the microscope several streams can be seen flowing in the surface layer parallel to the long axis of the stalk. The configuration of the bands is changing all the time. It is impossible to say whether these bands are paths with a definite structure, channels for the streams of cytoplasm, or whether they are devices which actively move the cytoplasm, but it is clear that they are essential for movement of the endoplasm.

The velocity of movement of the cytoplasm is of the order of 5 μ/sec but it varies slightly in different individuals, although it is constant in the same band. Addition of $5 \times 10^{-4} M$ ATP to the medium (sea water) soon increases the velocity of the streams by 4-5 times. The activated state lasts only about 10 min, after which, despite the presence of ATP in the medium, the velocity of movement returns to its initial value. ATP in higher concentrations, on the other hand, retards movement. For instance, after the addition of $2 \times 10^{-3} M$ ATP the cytoplasm moves more rapidly for 4-5 min, but then stops completely. ATP in high concentrations disrupts the bands, which simply disappear and are evidently gathered into lumps. The chloroplasts agglutinate and form conglomerates. Both the activating and the harmful action of ATP are specific for this substance, neither AMP nor sodium pyrophosphate or triphosphate have any action of this kind. Consequently, Acetabularia calyculus has a protein system which reacts specifically with ATP.

Takata also obtained glycerinated models from the stalks of this alga. The technique of their preparation is identical, in principle, to that of obtaining models from fibroblasts. The only difference is that the glycerol concentration in the extracting solution was reduced to 20%. The extracted cells retain the normal morphology of their internal contents although there is no movement of their cytoplasm. The addition of $1 \times 10^{-3} M$ ATP and $1 \times 10^{-3} M$ $CaCl_2$ or $MgCl_2$ to the preparation quickly causes movement of the protoplasmic gel. The movement does not last long, only 1-2 min. According to the graph given by Takata the velocity of this movement during the first 30 sec is 3 times greater than the velocity of movement in the normal living Acetabularia cell. It corresponds to that observed in the living cell when placed in ATP solution. Neither ATP itself nor inorganic phosphates can induce movement of the cytoplasm of the model: Ca^{++} or Mg^{++} must be present, besides ATP, for movement to arise. The ionic strength of the medium must be about 0.1; if greater than 0.3 no movement will arise. After brief movement the cytoplasm stops and, judging from the microscopic picture, it is in a state of more advanced coagulation than it was before addition of the ATP.

Takata concludes from these experiments that Acetabularia cells contain a protein sensitive to ATP, and that streaming of the cytoplasm is the result of interaction between this protein and ATP. This conclusion can evidently be extended to movement of the cytoplasm in other plant objects. Takata's findings also show that the mechanism of cell motility rests on a common basis in animals and plants. At this point it is worth recalling that actomyosin-like and myosin-like proteins have been obtained from the cells of certain plants. First Loewy (1952) and then Ts'o and co-workers (Ts'o et al., 1956a, 1957), obtained a protein with ATPase activity from the plasmodia of myxomycetes, which are distinguished by their vigorous cytoplasmic streaming. Ts'o and co-workers called the protein which they obtained myxomyosin, by analogy with actomyosin. The two share a number of very important properties, although there are also significant differences between them (for example, the change in viscosity of myxomyosin induced by ATP is not affected in any way by Ca^{++} and Mg^{++}), so much so that it has even been stated that myxomyosin is not a contractile protein (Hoffmann-Berling, 1960). However, a protein obtained by Nakajima (1960) from the same object by a somewhat different method (exactly the same method as is used to obtain actomyosin from muscles) is very similar to actomyosin. In particular, its ATPase activity is activated by calcium and magnesium ions.

Poglazov (1956) found ATPase in saline extracts of the leaves of several members of the acacia family. This ATPase has a pH optimum in the region 5-6 (myosin ATPase has two optima: at pH 9.0 and 6.3, and actomyosin has one — at pH 6.6), and its activity is not affected by Ca^{++} and Mg^{++}. It can also dephosphorylate ADP. Lyubimova and co-workers (1964) later discovered that the apyrase activity found by Poglazov is the result of the action of two enzymes: ATPase and ADPase, which can easily be separated. Unlike ADPase, ATPase is bound with the structural elements of the cytoplasm. In a later investigation Lyubimova and co-workers (1966) showed that the ATPase of mimosa leaves consists of two enzymes. One of these ATPases is found in the contracting parts of the mimosa — the cells of the pulvini of the leaf-stalks — and, contrary to Poglazov's (1956) findings, is activated by Ca^{++} and Mg^{++}, with a pH optimum at 8-8.5. In these last properties this ATPase resembles myosin. These workers also obtained contractile models from cells of the pulvini of mimosa leaf-stalks; after extraction in glycerol solution the tissue of the pulvini contracted in the presence of ATP and Mg^{++}.

A protein very similar to myosin has been isolated from Nitella by Vorob'eva and Poglazov (1963).

Kamitsubo (1969) found linearly oriented fibrils, about 1 μ apart, and parallel to the direction of the cytoplasmic streams, in the inner cells of the cortical layer of the nodes and rhizoids of Nitella and Chara. The cytoplasm streams very rapidly near them. If the streaming of the cytoplasm is stopped by electrical stimulation, it soons begins again, and certain unidentified particles in contact with the fibrils begin to move first. By centrifugation of cells or whole plants the fibrils can be separated from the cortex, and they then float freely inside the cells, in the cytoplasm. Some of them coil into loops, and these loops rotate on the spot and undulate in the stationary cytoplasm. Electrical stimulation stops them, and very soon after the cytoplasmic particles which are in direct contact with them begin to coil in a direction opposite to that of the preceding rotation of the loops, but then stop once again. After a short time the loop-shaped fibrils recover from the shock and resume their circular movement in the same direction as originally.

Kamitsubo's observations are very valuable. They suggest that active movement of the cytoplasmic streams in the plant cell is produced by the fibrils, by their undulation, propelling the passively moving cytoplasm with its organoids and inclusions. It is important to discover the nature of these fibrils, to compare them with the contractile protein structures of other objects, to investigate whether or not they possess ATPase properties, to discover the role of Ca^{++} and Mg^{++} in their functions, and to ascertain whether they have structures similar to the "arms" of the fibrils of flagella and cilia or the cross-bridges of myosin and paramyosin, i.e., whether they have suborganoids with which the ATPase activity and mechanochemical effect are directly associated. The solution to all these problems would give much valuable information on the molecular evolution of the mechanical function of protoplasm.

7. RHYTHMIC MOVEMENT OF FLAGELLA AND CILIA

Many hypotheses have been put forward to explain the rhythmic movements of flagella and cilia, but they are largely speculative. Electron-microscopic investigations have shown that all flagella and cilia of animals and plants are built according to the same plan, although with certain variations. Along the periphery of the flagellum or cilium run nine fibrils (in transverse section it is clear that they are arranged around the circumference, and equal distances apart, and are joined by cross-links or "arms." In the flagella and cilia of mammalian cells there are two concentrically arranged sets of these fibrils, with nine in each. In the center there are two fibrils, and movement of the flagellum (if two-dimensional) or cilium takes place as a rule in the plane perpendicular to that passing through the two central fibrils. The distribution of fibrils in flagella moving in one plane is the same as that of flagella moving in three dimensions.

One group of hypotheses (Gray, 1955) put forward to explain the continuity of flexion and extension of the cilium postulates that half of all its peripheral fibrils, namely the fibrils located on one side of the plane perpendicular to the axial plane which passes through the two central fibrils, bend simultaneously. Let us assume, for example, that all the left fibrils actively bend, in which case the cilium will bend to the left. Elastic forces then bring the cilium back to its central, neutral position, after which the right group of fibrils actively bends it to the right. The cilium thus oscillates like a swing. However, electron-microscopic observations provide no evidence for ascribing contractile properties to some structures and elastic properties to the others. It is only in the tails of mammalian spermatozoa that the fibril is surrounded by a helical structure which is considered to be an elastic, and not a contractile, element.

The same mechanism is postulated for three-dimensional movement. The basal body is the center of automatic rhythm, the pacemaker from which the excitation wave travels. It reaches the fibrils successively, so that the contraction of each successive fibril takes place a certain constant time after the contraction of its predecessor. This results in a helical movement.

Conflicting views are expressed regarding the actual role of fibrils of different types in flagellar movement; in particular, there is no agreement on the question of which fibrils conduct excitation. A detailed analysis of this problem, as of others to do with ciliary and flagellar movement, can be found in the book by Seravin (1967) and the surveys by Gibbons (1967) and Brokaw (1968). It is very important to note that, as direct measurements have shown, during bending of the cilium the fibrils do not shorten.

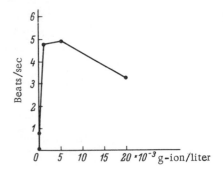

Fig. 36. Frequency of beating of model flagella of locust spermatozoa as a function of Mg^{++} concentration (from Hoffmann-Berling, 1955). Abscissa, Mg^{++} concentration; ordinate, frequency of beating of model flagella. Points: A) in medium without addition of Mg^{++}, B) with addition of 5×10^{-3} M EDTA. Models obtained by extraction for 48 h. Reactivating solution: 1×10^{-4} M ATP, $\mu = 0.20$, pH 6.8.

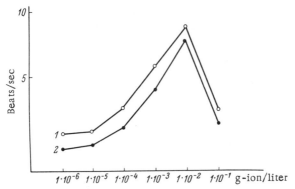

Fig. 37. Frequency of beating of model flagella from spermatozoa of the sea urchin (1) and starfish (2) as functions of Mg^{++} concentration (after Kinoshita, 1958). Axes as in Fig. 36. Reactivating solution: 1×10^{-3} M ATP, 1×10^{-2} M histidine, pH 7.5.

This suggests that bending is based on a mechanism of sliding of the mechanically active elements inside the fibril along each other and not a mechanism of shortening.

Experiments by Hoffmann-Berling (1955) demonstrated the ATPase nature of flagellar movement. By glycerol extraction he was able to obtain models of the tails of spermatozoa of the tropical locust Tachycines, as well as of blowflies, cocks, and mice. These models perform rhythmic movements in the presence of ATP. Mg ions are necessary in very small quantities, but in their absence no movement takes place (Figs. 36 and 37). Versene, by removing Mg^{++}, inhibits the motility of the model tails. Motility can be restored by magnesium but not by calcium. The sulfhydryl poison Salyrgan inhibits the activity of these models, and monoiodoacetate has a weak action. Polysulfonic acids, such as Germanin and heparin, likewise inhibit movement.* Admittedly, in models subjected to prolonged extraction, which have lost their elasticity, polysulfonic acids act differently: by plasticizing the models they

---

*The possibility cannot be ruled out that this is of great biological importance. On penetration of the spermatozoan into the ovum it forms a fertilization membrane by secreting heparin-like polysulfonic acids. This may be the way in which movement of other spermatozoa, attempting to penetrate into the ovum, is arrested so that polyspermia is prevented (see also page 88).

improve their function. On the whole, however, the inhibitor pharmacology is the same for these models as for muscle models. ATP in high concentrations has a superoptimal effect on model flagella as on other models: 2×10^{-2} M ATP does not induce their movement. The magnitude of the superoptimal ATP concentration is reduced if the ionic strength of the solution is increased — this is also analogous to what is observed in muscle models. The above remarks are evidence that ATP provides the source of energy for flagellar movement. ATP is specific and cannot be replaced by other organic phosphates or by inorganic polyphosphates. ATP undoubtedly induces the beating of model flagella by reacting directly with the contractile proteins. Tails of the giant spermatozoa of <u>Tachycines</u> consist of three fibrils coiled one around the other. In model tails they can be separated by the action of weak alkaline solution, and each fibril functions in the presence of ATP. If the model tail is separated from the basal body, it will then perform rhythmic movements under the influence of ATP and Mg^{++}. Furthermore, the model flagellum can be cut transversely at its center and both halves will be capable of functioning. Only the very tip, the distal third of the flagellum, if separated is unable to move: it probably has no contractile elements either in the living or in the model flagellum, and it moves passively.

Model tails can oscillate only in one plane. Such models cannot perform threee-dimensional rotation or a wave of contraction along the flagellum. The frequency of beating of model flagella is always lower than that of living spermatozoa. In ATP in concentrations of between 1×10^{-5} and 1×10^{-2} M it is directly proportional to log [ATP] (Fig. 38, curves 1 and 2; Fig. 39). The amplitude of the beats of model flagella is inversely proportional to their frequency (Fig. 23, curve 3). All factors which increase frequency at the same time reduce amplitude. The frequency of beating can be increased by means of agents plasticizing the muscle model, including inorganic pyrophosphate and urea, and also by increasing the ionic strength of the solution against the background of ATP in minimal concentration (6×10^{-6} M) inducing movement (Fig. 39). Protamine sulfate, on the contrary, "strengthens" the models and makes them more rigid, thereby reducing the frequency of beating. It will be recalled that protamine sulfate inhibits relaxation of connective-tissue cell models (see p. 73).

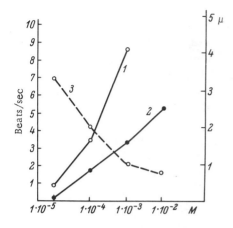

Fig. 38. Frequency (1 and 2) and amplitude (3) of beating of model flagella from locust spermatozoa as functions of ATP concentration (from Hoffmann-Berling, 1955). Abscissa, ATP concentration; ordinate: on left, frequency of beats of model flagella, on right, amplitude of beats. 1 and 3) Models obtained by extraction for 2 days, 2) models obtained by extraction for 16 days. Reactivating solution: $\mu = 0.20$, pH 7.2.

The strength of the beats is reduced if the models are kept in the extracting solution as they "grow old." Models extracted for a short time only are able to make propulsive movements, while those kept for a long time demonstrate only the oscillatory movements of the flagellum and do not move from their place. In "young" models the same phenomenon is observed if they are placed in a viscous medium.

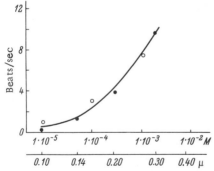

Fig. 39. Frequency of beats of model flagella of locust spermatozoa as a function of ATP concentration ($\mu = 0.20$, circles) and of ionic strength (5×10^{-4} M ATP, points) of reactivating solution (from Hoffmann-Berling, 1959). Abscissa: top) ATP concentration, bottom) ionic strength (in relative units); ordinate, frequency of beats of model flagella.

In briefly extracted models of unicellular Flagellata (Peranema trichophorum) ATP causes waves to spread along the flagellum, while in models extracted for a longer time it induces only stationary waves (Seravin, 1967). Waves of beating always arise in Peranema at the base of the model flagellum and spread toward its end. In model flagella separated from the cell body and cut into quarters, waves also spread from the proximal part toward the distal. Potassium thiocyanate and potassium iodide, in concentrations of 0.3-0.5 M, which (as has already been mentioned) can induce contraction of a muscle model in the absence of ATP (Laki and Bowen, 1955; Laki, 1967), initiate stationary waves of model Peranema flagella. Thiourea, which plasticizes muscle models, produces straightening of bent model flagella in a concentration of 0.2-0.3 M in the absence of ATP.

The temperature dependence of the frequency of beating of model spermatozoal tails is the same as that of the rate of contraction of muscle models: within the range 0-20°C, Q_{10} = 2-3 (Fig. 40). At pH values below 7.0 and above 8.1, model flagella are unable to move; the optimum pH lies within 7.3 and 7.5 (Fig. 41).

It will be recalled that with other types of models a decrease in ionic strength or an increase in acidity of the reactivating solution inhibits relaxation and promotes contraction. In the case of model flagella, on the other hand, it is impossible in either way to obtain contraction in isolation from relaxation. This makes it impossible to say which phase of the cycle requires ATP. It may be supposed that ATP is hydrolyzed, as in muscle, in connection with contraction, but it is equally possible to make the opposite suggestion: just as in the Vorticella stalk, ATP induces relaxation.

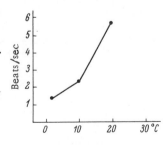

Fig. 40. Frequency of beats of model flagella of locust spermatozoa as a function of temperature (from Hoffmann-Berling, 1955). Abscissa, temperature of reactivating solution; ordinate; frequency of beats of model flagella. Mean value of Q_{10} is of the order of 2. Reactivating solution: 1×10^{-3} ATP, μ = 0.18, pH 7.2.

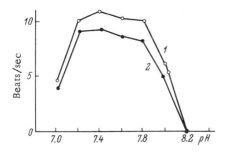

Fig. 41. Frequency of beats of model spermatozoal flagella of sea urchin (1) and starfish (2) as a function of pH of reactivating solution (from Kinoshita, 1958). Abscissa, pH; ordinate, frequency of beats of model flagella. Reactivating solution: 1×10^{-3} M ATP, 5×10^{-3} g-ion/liter Mg^{++}, 1×10^{-2} M histidine.

On a priori grounds there is even a third possibility, namely that ATP is essential for both phases of the cycle of flagellar movement, acting in one phase, for example, as the source of energy and in the other phase as a plasticizer. However, this is all pure speculation.

Bishop (1958c) showed that a preparation of relaxing granules obtained from rabbit muscles inhibits the movements of model spermatozoal flagella unless the solution contains an excess of Ca^{++}. Since under these conditions the first phase (bending of the flagellum) disappears, it would seem that the contractile movement is suppressed. However, this tells us nothing about whether ATP is required for contraction or for relaxation, because relaxing factor in the case of models of muscles and fibroblasts inhibits contractility induced by calcium without the participation of ATP.

The inhibitory action of relaxing factor on flagellar models can be abolished, as in other cases, by calcium.

The list of animals from whose spermatozoa models have been made has been considerably lengthened after the appearance of Hoffmann-Berling's fundamental paper (Hoffmann-Berling, 1955). Models have been made from spermatozoa of the cuttlefish (Bishop, 1958b), starfish and sea urchin (Kinoshita, 1958), rabbit, guinea pig, rat, ox, and man (Bishop and Hoffmann-Berling, 1959), and frog (Ivanov et al., 1959).

A special feature of model spermatozoa of the starfish and sea urchin is that they function only if, besides ATP and Mg^{++}, the medium contains a chelating agent (histidine, cysteine, thioglycollate, versene). It can be replaced by a preparation of relaxation granules from vertebrate muscles or by the analogous factor obtained from the spermatozoa of the starfish or sea urchin themselves. This agent is physiological in character, probably with a role in the normal function of the flagella (Kinoshita, 1959).

Hoffmann-Berling obtained similar models from trypanosomes in which, after glycerol extraction and on addition of ATP and Mg^{++}, the flagellum and undulating membranelle make rhythmic movements. Everything which was said above about model flagella from spermatozoa applies also, in general, to model flagella of trypanosomes. As a minor point of difference, the superoptimal effect — inhibition of function of the models by ATP in high concentrations — is induced in trypanosomal models by lower concentrations of ATP, in fact as low as 5×10^{-3} M. Their sensitivity to polysulfonic acids and, in particular, to Germanin, is higher. Germanin has also been suggested as a suitable drug for the control of trypanosomal infections. Finally, model flagella of spermatozoa can withstand an increase in ionic strength only up to 0.4, compared with 1.0 for trypanosomes.

In investigations of great theoretical importance Brokaw obtained glycerinated models of flagella of plant objects — the unicellular algae Chlamydomonas (Brokaw, 1960) and Polytoma uvella (Brokaw, 1960, 1961, 1963, 1966), which were activated by ATP and Mg^{++}.

This shows that flagellar movement is based on the same principle of energetics in the animal and plant worlds. The fact that the mechanism of movement is the same in both cases is also indicated by the similarity between the electron-microscopic pictures of flagellar structure in animal and plant cells.

One difference between models of the flagella of lower plants and models of animal flagella is that the plant models can perform three-dimensional movements and, because of this, they can move directively.

Models have been obtained (Aleksandrov and Arronet, 1956; Arronet, 1964b, 1968) from cilia of the ciliated epithelium of the palatal mucous membrane of frogs and toads, the lower part of the frog oviduct, and the mouse and rabbit trachea. Models of ciliated epithelium from the frog palate have been studied most fully (Fig. 42). The tissue undergoes maceration on glycerol extraction, and it is therefore convenient to examine scrapings from

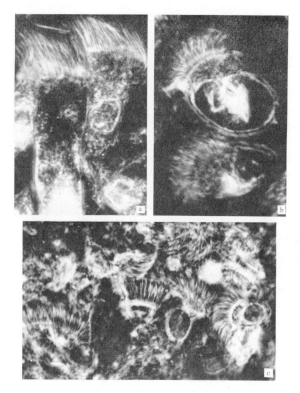

Fig. 42. Glycerinated models of ciliated epithelial cells from the palate of <u>Rana ridibunda</u> (original photographs). (a) and (b) anoptral (phase-contrast) microscope, 720×; (c) phase-contrast microscope, 320×, extraction for 48 h. Cytoplasm and contents of nuclei have coagulated.

the palatal mucosa under the microscope. The scraping consists of a suspension of groups of cells, individual ciliated cells and their fragments, and isolated cilia. The cilia are nonmotile, the cytoplasm of the cells is coagulated, the models stain diffusely with vital dyes like all dead cells, they fluoresce brightly in a dark field, and after staining with fluorochromes they fluoresce like dead cells. The cells are, in fact, dead. However, the contractile system of their cilia remains intact: if ATP (in concentrations of the order of $5 \times 10^{-3}\ M$) and $5 \times 10^{-3}\ M$ $MgCl_2$ are introduced under the cover slip the cilia of most cells (groups of cells, individual cells, fragments of their apical parts sometimes consisting only of the cell membrane carrying the cilia and, what is particularly important, detached, isolated cilia suspended in the solution) begin to oscillate energetically. After a considerable time, measured in tens of minutes, the oscillation slows down and then ceases altogether.

It can be restored by introducing a fresh portion of ATP, indicating that ATP is used up in the course of the oscillatory movement. At room temperature, if ATP solution is added from time to time to the preparation, oscillation can be maintained for about 2 h. The minimal ATP concentrations causing oscillation of these models are of the order of 1×10^{-4} M. With an increase in concentration the frequency of beating of the cilia rises and their amplitude falls. ATP in a concentration of 1×10^{-2} M causes oscillation of such high frequency and low amplitude that it gives the impression that the cilia are vibrating. Concentrations above 2×10^{-2} M are superoptimal. Magnesium ions are essential for oscillation. Other phosphates, such as glycerophosphate and pyrophosphate, cannot replace ATP in its role of the source of energy for oscillation. Normal coordination between the oscillations of the cilia is not maintained over the whole cell. The cilia either move completely at random, each completely by itself, or they bend and straighten in unison, which is never seen in the living cells.

Cells of the "uterine" part of the frog oviduct do not undergo maceration on extraction, so that the whole sheet of model cells is examined under the microscope. The fact that the model of ciliated cells of the oviduct is a single sheet of tissue makes it convenient for work requiring the transfer of models from one solution to another, such as is the case when the action of various substances on them must be tested; under these circumstances the model cells do not fall off the sheet and there is no risk of their being crushed (Arronet, 1968). Examination of the movements of model oviductal ciliated cells shows that the coordination of their activity in the sheet is disturbed.

Coordinated function of the cilia is preserved in the models of Paramecium which the writer has observed, and as a result the models of these infusorians in ATP + Ca^{++} solution can move about slowly.* Peristomal membranelles of Euplotes patella models

---

*Seravin (1961) considers that priority in the obtaining of ciliary models from infusorians belongs to Aleksandrov and the present writer. This, in fact, is not the case. Models of Paramecium were obtained originally in 1957 by Hoffmann-Berling, and

work together in relative unison, so that these models also can move about (Seravin, 1967). An extremely interesting fact has recently been discovered by Naitoh and Jasumasu (Naitoh, 1966; Naitoh and Jasumasu, 1967). They observed reversal of ciliary oscillations in models of Paramecium.†

The writer has found that models of Paramecium prepared by digitonin extraction can still form deposits of granules of the vital dye neutral red.

Satir and Child (1963) examined model cilia prepared by glycerol extraction from the ciliated epithelium of two species of frogs under the electron microscope. The living, unextracted cilium contains a complex of 9 + 2 fibrils, surrounded by an amorphous substance and bounded externally by a membrane, as has repeatedly been described. During preparation of the models the outer membrane of the cilia is disrupted into fragments and dissolves, and the material of the structureless outer zone swells considerably. The filament complex is exposed and also swells, and after very prolonged extraction its nine external fibrils can actually be displaced. Satir and Child consider that in those model cilia which are capable of reactivation, the membranes are destroyed, the matrix and fibrillary material are extracted, but the fibrils of the 9 + 2 complex themselves remain intact and are not displaced. If they had been displaced, the model cilium would no longer be able to respond to ATP by oscillation.

Model cilia have also been obtained from ciliated cells of the branchial plates of the mussel Modyolus demissus (Child and Tamm, 1963). In these models they observed uncoordinated movements, synchronized movements (within the same cell), and wavelike spreading of movements over all the cilia of a given cell.

Gibbons (1965) obtained and investigated model cilia from the infusorian Tetrahymena pyriformis. He ground the cells in a

---

I prepared models from Spirostomum (with the aid of digitonin and saponin extraction) and from Paramecium caudatum (by digitonin extraction) in 1958, using the directions given to me in a letter by Hoffmann-Berling. In 1961, Seravin also prepared models of infusorians — from Spirostomum ambiguum as before and, in addition, from Euplotes patella. Having obtained models from Euplotes patella, Seravin showed that movements of the cirri are based on the same biochemical mechanism as movements of cilia, flagella, and undulating membranes.
†See Notes Added in Proof, page 172.

Vortex mixer, then centrifuged the suspension and isolated a suspension of cilia. These isolated glycerol-extracted cilia can be reactivated by a solution containing ATP and Mg^{++}, in which they oscillate energetically. Several types of oscillations are observed. The commonest are undulating in type: the waves travel along the cilium from its basal end at a frequency of 2-3 waves per second; the cilia usually turn on their axis and also advance with a twisting motion, the basal end foremost, at a speed of about 2 μ/sec. The other cilia do not advance but simply bend rhythmically in various ways and oscillate, and can be described as being of the beating type.

Immediately after model cilia are placed in reactivating solution the proportion of motile cilia, according to Gibbons' calculations, is about 80%. This falls to 30% after 5 min and to 1% after 15 min. In most cilia 5 min after beginning of reactivation the undulating movement changes into movement of the beating type. After this conversion or degradation of the movement the cilium frequently disintegrates or curves into a ring. Inactivation of the cilia is probably not due to thermal injury to their proteins: at 0°C the duration of movement is no longer than at room temperature.

Variations in concentrations of ATP and Mg^{++} and in the ionic strength of the reactivating solution affect the frequency of oscillation of the cilia, and the character of the effects is the same as those observed by Brokaw (1961) in his experiments on model flagella of Polytoma. Reactivation is possible at pH values between 6.2 and 8.5, but the duration of ciliary movement reaches a maximum at pH 6.8. After the models have been kept in extracting solution the proportion of cilia capable of reactivation decreases.

The following fact appears significant: chelates only slightly inhibit the motility of model cilia, and their action is the same at pH 6.8 and at pH 7.8, which is not typical of the chelate effect; furthermore, this weak inhibitory action is not abolished by the addition of 1×10^{-3} M $CaCl_2$. Gibbons (1965) calculated that movement of model cilia in the presence of ATP and Mg^{++} is not inhibited as a result of a decrease in the Ca^{++} concentration at pH 7.8 to 1×10^{-9} g-ion/liter. This distinguishes the cilium sharply from muscle: lowering the Ca^{++} concentration to 5×10^{-7} g-ion per liter completely abolishes the contractions of model myofibrils and also the sineresis of the actomyosin gel under the influence of

ATP. This different behavior toward calcium ions is interpreted by Gibbons as an important difference between the motile mechanism of muscles and cilia.

Gibbons describes some characteristic features of the movement of single model cilia from Tetrahymena pyriformis. First, the movement is helical and not two-dimensional as in model flagella. Second, movement of the cilia, which caused the basal body to adhere to the slide, in all cases (about 100) observed took place clockwise; this evidently reflects certain features of the ciliary movement of the normal cell.

Winicur (1967) used the ethanol-calcium treatment of Watson and Hopkins (1962) instead of mechanical separation of the cilia from the Tetrahymena cells and also obtained isolated cilia which were then extracted, for which purpose several solutions were tested (containing either glycerol in various concentrations or 0.05% digitonin with 60% sucrose, or ethylene glycol), and he prepared models from the cilia. These model cilia can be reactivated not only by ATP, but also by ADP. However, oscillations induced by ADP arise after a short latent period: 20-30 sec if $3 \times 10^{-4} M$ ADP is used, and somewhat shorter if $3 \times 10^{-3} M$ ADP is used. This agrees with the observations of Brokaw (1961), made on model flagella of the unicellular alga Polytoma uvella. Winicur considers that ADP does not itself induce movement of the model cilia, but is first converted into ATP by the adenylate kinase present in the cilium.

The fact that model flagella and cilia contract and relax under the same conditions as those in which muscle models contract and relax indicates that the mechanisms of movement are based on common principles in both cases.

Further evidence of the similarity between the mechanisms of movement of flagella and cilia, on the one hand, and muscles on the other hand, was provided by the isolation of contractile proteins resembling muscle protein from flagella and cilia. The investigations of Éngel'gardt and Burnasheva led to the isolation of a contractile protein called spermazine from bovine spermatozoa Éngel'gardt, 1957; Éngel'gardt and Burnasheva, 1957). It is localized in the tails of the spermatozoa and can be activated by calcium ions. Bishop (1958a) also obtained a protein from bovine spermatozoa which he regards as a component of a contractile protein complex analogous to myosin. He suggests that a second component also exists, and that this is analogous to actin. However, the myosin-like protein could not be combined with muscle actin. The protein which Bishop obtained gives superprecipitation in the presence of ATP and Mg^{++}. A protein consisting of two components, one resembling myosin and the other resembling actin, has been obtained by Nelson (1966) from the spermatozoa of certain marine in vertebrates.

A contractile protein with ATPase activity has also been obtained from isolated cilia of the infusorian Tetrahymena pyriformis (Burnasheva et al., 1964; Gibbons and Rowe, 1965). The ATPase activity of this protein, known as dynein, is increased by calcium ions and, to a lesser degree, by magnesium. This ATPase is bound with the structural elements of the cilia themselves, it is specific toward ATP, and it is inhibited by versene. Gibbons (1967) showed that the protein with ATPase properties is localized in the "arms" of the cilia — the submicroscopic projections linking the peripheral intraciliary fibrils. According to Burnasheva et al. (1969), both the central and the peripheral fibrils of cilia and flagella consist of globular subunits. ATPase is associated with the subunits composing the walls of the various fibrils, and it is also found in the matrix.

* * *

Experiments on the various types of contractile models have thus revealed four types of motile mechanisms. One type consists of contraction induced by ATP and associated with its hydrolysis; it is by this mechanism that muscles, connective-tissue epidermal cells, amebas, mitochondria, chloroplasts, the chromosome threads of the mitotic spindle, and structures forming the equatorial constriction of the cell during its division contract. The second mechanism is one of elongation as a result of ATP binding (examples are the relaxation of muscle and lengthening of the central spindle and body of the cell in anaphase). The third mechanism is contraction by the action of Ca^{++} and elongation by the action of ATP (the Vorticella stalk). The fourth mechanism is elongation by the action of Ca^{++}: this mechanism is prevented from operating by ATP (trichocysts).

Clearly, therefore, ATP is directly concerned in the work of three of these mechanisms; it can induce either contraction or lengthening of motile structures. In a class by itself is **lengthening of trichocysts,** in which the role of ATP is merely that of inhibiting lengthening induced by calcium. The most widespread mechanism of contraction is that related to contraction of the cross-striated muscle, i.e., enzymic interaction between the contractile enzyme-protein and ATP as the energy-providing substrate. Movement of flagella and cilia is also associated with hydrolysis of ATP, but it is difficult to say when the ATP is utilized: only in the contraction phase, or during relaxation as well. So far as movement of the cytoplasm in plant cells is concerned, it is now clear that its source of energy is also ATP, but the precise details of how ATP is utilized have not yet been discovered.

It is now becoming increasingly clear that the calcium mechanism of motility is very widespread and that Ca^{++} plays an essential role in the initiation and coordination of movement. In this context the work of Naitoh and Jasumasu (Naitoh, 1966, 1969; Naitoh and Jasumasu, 1967) to which reference has already been made, which demonstrated the leading role of Ca^{++} in the reversal of ciliary oscillations in Paramecium, is particularly important. Zn^{++} ions are also essential for this to take place. Facts of this nature continue to accumulate.

Chapter III

The Use of Motile Models to Study Problems of a Nonmechanochemical Nature

The chief field of application of contractile (motile) models is the investigation of the mechanism of muscle movement and of cell movement in general. It was for this purpose that models were originally suggested. The results obtained in this field by means of models have been described in the previous chapters. We can now turn to certain other spheres where models have been used, and where they have also proved valuable as a research tool.

Whenever some special feature is discovered in the motor activity of a cell, some change in its parameters or disturbance of its function, it must be asked whether changes in the contractile system of the cell itself or in some other system on which the act of contraction depends are responsible for the phenomenon observed. Models can be helpful in the study of this problem.

If a certain aspect of the motor activity of a particular object is reproduced in the work of a model obtained from it, it can be considered that this aspect is characteristic of the actual mechanically active system of that object. Furthermore, if a certain change in the behavior of a living contractile cell is also visible in the model prepared from it, the change can naturally be ascribed to a change in the contractile system itself. If, however, this change is not found in the model, it must be looked for not in the contractile system, but in other systems of the cell connected with it.

Investigations have been carried out on glycerinated muscle fibers in order to find out which protein substrate is responsible for the changes in sorption of vital dyes by muscle which arise after application of various stimuli. Suzdal'skaya (1964) showed that under the influence of various stimuli the affinity of muscle tissue for dyes changes in different ways depending on whether the stimulus used is adequate or inadequate. From this starting point, Ovsyanko (1968) and Suzdal'skaya and Troshina (1968) set out to discover what changes take place in the ability of the protein contractile system of muscles to stain with vital dyes after adequate and inadequate stimulation. As their test object they chose glycerinated fibers of the rabbit psoas muscle, and as the stimuli they chose ATP and heat (Ovsyanko, to 50-80°, time not specified; Suzdal'skaya and Troshina, to 80° for 1 min). Both ATP and a high temperature caused the models to contract, but the action of ATP in this case was that of a specific, adequate agent, while heat was nonspecific and inadequate. The sorptive capacity of the glycerinated fibers was found to change in different ways depending on the factor inducing contraction. These workers accordingly conclude that the mechanisms of contraction of the model in response to ATP and to heat are different. This was the conclusion reached previously by Portzehl (1951) from her experiments with actomyosin threads (see page 46 and Fig. 10).

From their comparison of the reaction of model fibers to ATP (adequate stimulus) and to heat (inadequate stimulus) and the reaction of living muscles to electrical stimulation (adequate stimulus) and to heat (inadequate), Suzdal'skaya and Troshina (1968, page 1537) reached the following conclusion: "the staining reaction of living muscles in glycerinated models is reproduced to inadequate stimuli but is not reproduced to adequate stimuli."

Investigations of this character are preferably carried out on glycerinated myofibrils or on fibers extracted with glycerol and detergent, for ordinary glycerinated fibers contain, besides contractile proteins, other protein components which can also take up dyes and in which this property could conceivably be altered by stimulation. In this particular case, the workers concerned were interested in changes in the sorptive properties of the contractile structures only.

Investigations into the causes of disturbance of cell contractile activity under the action of different agents has been carried out with the use of models. Brokaw (1960), for instance, investigated a mutant of Chlamydomonas moewusii which is nonmotile although it has flagella. What is the reason for the nonmotility of these cells? Can it be injury to the contractile system of the flagella or disturbance of accessory systems? The reason for inability of the flagella to function could be a disturbance either of the ATP-generating system or of the system responsible for excitation and its spread, or of the mechanism of ATP transportation, for example, from the mitochondria to the contractile elements, and so on. Injury to any one of these components could give the same result: inactivity of the flagella. So far neither light nor electron microscopy has revealed any differences in the morphology of the flagella of the normal mutant cells; all that has been found is that occasionally flagella of the mutants are slightly shorter than those of the normal cells. Likewise, no serological differences have been found between the motile and nonmotile flagella. Brokaw made models from the motile and nonmotile cells. The flagella of the models from normal cells were reactivated extremely well by a solution containing ATP. The flagella of models of mutant cells did not move in this solution. Brokaw concluded that the reason for the nonmotility of the mutant is damage to the contractile system of the flagella. He found, moreover, that the flagella of the nonmotile mutant cells have sharply reduced ATPase activity.

Muscle models have similarly found application in medicine: they have been used to analyze the character of muscle damage in certain diseases. Ivanov and Yur'ev (1961) described the results of their observations on the contractile ability of glycerinated models of fibers taken from muscles of patients with the sequelae of poliomyelitis. The degree of contraction induced by ATP in these model fibers was significantly reduced: some of them contracted by only 10-25% of their initial length. No actomyosin threads could be obtained from extracts of these muscles, probably because of the low actomyosin content in the extract (and also in the muscle). Consequently, the pathology of the muscle in this case is directly connected with injury to its contractile elements.

The writer has used models to identify the system which is damaged in the cell by the action of heat, thus inactivating the cilia on ciliated epithelial cells of frogs (Arronet, 1964b). Pieces of tissue from the ciliated epithelium of the palate or esophagus were heated for 5 min at temperatures differing from each other by steps of 0.2°C. After a certain minimal temperature, not a single oscillating cilium was left in the pieces of tissue. This temperature was used as a measure of the resistance of the cells to heating for 5 min. The resistance of ciliated epithelial cells to heat determined in this way is relatively constant for each species (Aleksandrov, 1952), although it does undergo regular changes in the course of the year (Dregol'skaya, 1963). The maximal temperature to which ciliated epithelial cells could be heated for 5 min and models reactivated in ATP solution still be obtained from them also was determined. This was done as follows. Pieces of frog epithelium were heated to various temperatures differing from each other by 0.4°C. After heating they were immersed in the extracting solution, and after extraction for 24 h tests were carried out to determine whether the models obtained from the cells of each batch after heating could be reactivated. In this way it was possible to discover whether the contractile system of the cilium was intact or had been damaged by heat at the given temperature; thus the limit of resistance of the contractile system to heat was established.

The results showed that it is impossible to obtain reactivatable models from cells whose cilia have been immobilized by heat. Conversely, if some cells with oscillating cilia still remained after heating, models with cilia reactivated by ATP can be obtained from them.

If the figures characterizing the resistance of ciliary function of living cells and the resistance of the contractile system (the heat resistance of the "ability of the cells to yield reactivatable models") are compared it will be clear that these figures are very close in every case — i.e., frogs of different species and the ciliated epithelium obtained from different organs (Table 2). Meanwhile, the heat resistance of the ability of cells to yield models is always slightly lower than the resistance of ciliary function of the living cells. This can easily be understood: the extraction process, i.e., preparation of the models, itself damages the contractile system to a certain extent and it is perfectly natural that this damage is sufficient to tip the balance, for cells subjected to intensive heating, almost strong enough to suppress ciliary movement completely are on the borderline of injury. The following facts are evidence of the injurious effects of extraction on the contractile system of ciliary movement. If models are kept in extracting solution in the cold

TABLE 2. Heat Resistance of Ciliated Epithelial Cells of the Frog Palate and Oviduct and of the Ability of the Cells to Yield Models

Object	Species of frog	Temperature (in °C) to which tissues must be heated for 5 min to suppress	
		ciliary function of living cells	ability of cells to yield models
Ciliated epithelium			
Palate	Rana ridibunda	44.4	43.7
	Caucasian frog (Rana camerani)	42.8	42.6
	Rana temporaria	41.9	40.9
Oviduct	Rana ridibunda	44.5	43.7
	Caucasian frog (Rana camerani)	42.3	41.5
	Rana temporaria	41.5	40.8

their resistance to heat falls steadily: it is lower after extraction for 3 days than for 1 day, and it is lower after 7 days than after 3 days (Aleksandrov and Arronet, 1956). With an increase in the duration of extraction the resistance of models to the action of a high hydrostatic pressure also falls (Arronet and Konstantinova, 1964).

Although it is difficult to find cells with nonfunctioning cilia in a fragment of ciliated epithelium recently taken from an animal, nevertheless after extraction for the optimal time nothing like 100% of the model cells of the preparation will react. It is likely that the contractile system of the cilia of some of the cells is injured during extraction.

A similarity was thus detected between the values obtained for the thermostability of the ciliary function of living cells and of the contractile apparatus (the thermostability of the ability of the cells to yield models) of cilia taken from frog ciliated cells. This has been demonstrated for ciliated cells from different sites (palate and oviduct) in 3 species of frogs. Hence it was concluded that it is the level of thermostability of the contractile ciliary apparatus which limits and defines the thermostability of the ciliary function of living cells. In other words, the cause of inactivation of the cilia of cilated cells exposed to the action of heat is injury to the ciliary contractile apparatus. Further analysis of thermal injury to the contractile apparatus of ciliated cells revealed that it consists of thermal denaturation of proteins composing the contractile system. This is shown by the high temperature coefficient of thermal inactivation of model ciliated cells: in the case of model cells from Rana temporaria $Q_{10} = 35$ (Arronet, 1964a).

It would seem that the thermostability of the contractile system can also be determined in another way: by exposing not living cells, but models made from them, to the injurious action of heat. Experiments of this type have also been carried out, but it has been necessary to apply the heat to a contractile apparatus in a nonphysiological environment, for the remaining components of the cell which normally surround the contractile apparatus are absent from models: they have been removed by extraction. It is therefore not surprising that the thermostability models of ciliated epithelium has been found to be below the thermostability of the ability of the cells to yield models, for when this latter parameter is determined the contractile system of the cilia during exposure to heat still remains in the cell; it is lower than the thermostability of ciliary function of living cells by an even greater margin.

Models have been used similarly to analyze the injurious action of a high hydrostatic pressure on ciliated epithelial cells (Arronet, 1964a; Arronet and Konstantinova, 1964). In ciliated epithelial cells exposed to this agent in a certain intensity, the ciliary function ceases. Just as in the case of thermal inactivation, it is impossible to obtain models reactivable by ATP from cells exposed to pressure sufficiently high to suppress ciliary motility. Conversely, working models can be obtained from cells exposed to pressure in intensities not sufficient to suppress ciliary function completely. The values for the resistance of ciliary function of living cells and their ability to yield models to increased hydrostatic pressure are very similar, and for palatal cells of Rana temporaria they are 1590 and 1560 atm respectively for an exposure of 5 min. If previously prepared models are exposed to a high hydrostatic pressure, and their resistance to this agent is determined in the same way, a figure very close to the limit of resistance mentioned above is obtained, namely 1500 atm. It is demonstrative that in the case of Rana ridibunda, whose cells are more resistant to the action of hydrostatic pressure than those of R. temporaria, models of these cells are also more resistant to this agent. For instance, with an exposure of 5 min the limit of resistance of ciliary function of living cells is 1690 atm, compared with 1650 atm for models.

We can conclude from these facts that the injurious action of hydrostatic pressure on ciliary function is based on injury to the contractile apparatus of the cilium by this agent.

It has been shown by the use of models that the injurious action of these agents — heating and hydrostatic pressure — on muscle rests on a different basis (Arronet and Konstantinova, 1969). Each of these agents, when applied in a certain dose, renders the muscle unable to respond by contraction to direct electrical stimulation. However, the contractile system of such a muscle, although it does not respond to electrical stimulation, may still be intact: after exposure to pressure or heating in a dose much higher than the smallest dose which inhibits electrical reactivity of the muscle, it is still possible to obtain a glycerinated model fiber from it which will contract in reactivating solution. The resistance of the living muscle to both these agents, if estimated from its electrical reactivity, is significantly lower than the resistance of the contractile system determined either from its "ability to yield models" (treatment of the living fiber, followed by preparation of a model from it and assay of its ability to respond to ATP) and from the resistance of the model fiber (treatment of a previously prepared glycerinated fiber). Evidence to support this statement is given by the figures in Table 3. It is demonstrative that the limit of resistance of the model fibers or of the "ability of the muscles to yield models" to the action of pressure is virtually the same as the dose of that agent which completely suppresses the ATPase activity of actomyosin (Arronet and Konstantinova, 1969).

It was concluded from these facts that under the action of a high hydrostatic pressure or of heating the muscle loses its ability to respond to electrical stimulation by contraction, as the result of injury not to its contractile apparatus, but to some other system probably connected with excitation. This conclusion drawn by the present writers agrees with that which Ushakov (1963, 1965, 1967) drew regarding the action of heat on muscle on the basis of experiments of a different type.

TABLE 3. Resistance of Muscle Fibers of Rana temporaria and of Their Contractile Apparatus to Exposure for 15 min to Hydrostatic Pressure or to Heating

Muscle fibers	Level of resistance of		
	living fibers	ability of fibers to yield models	models
		Hydrostatic pressure (in atm)	
M. sartorius	700	2550 ± 119	2850 ± 188
M. ileo-fibularis	700	2580 ± 58	2477 ± 182
		Heating (temperature, in °C)	
M. sartorius	37.1 ± 0.5	38.4 ± 0.5	38.7 ± 0.3
M. ileo-fibularis	37.3 ± 0.4	38.1 ± 0.24	38.4 ± 0.15

The point of application of the injurious action of heat and of a high hydrostatic pressure on muscle thus differs from its point of application on ciliated epithelial cells. In the latter, as has already been stated, these factors suppress motile function by injury to the contractile apparatus, which is not observed in muscles. Admittedly, we must qualify this remark by saying that the degree of injury to the muscles is judged by their loss of electrical reactivity, while that of ciliated epithelial cells is judged from the cessation of spontaneous motility of their cilia. These two criteria, of course, are not identical.

A case has been described in which the cessation of ciliary function in ciliated epithelium is not due to injury to the contractile apparatus, as was shown by the fact that highly reactivatable models can be obtained from cells with nonfunctioning cilia (Arronet, 1968).

Movements of cilia of the oviductal ciliated cells of Rana temporaria can be stopped by placing them for 20 min in $5 \times 10^{-3} M$ NaN_3 or for 30 min in $5 \times 10^{-2} M$ NaF made up in Ringer's solution. However, models obtained from cells with cilia inactivated by these metabolic inhibitors can be reactivated and will continue to work for a long time in a solution containing ATP, Mg^{++}, and the inhibitor in the same concentration as that in which it abolished the motility of living cilia. In these experiments the appropriate inhibitor was added to the extracting solution also so that the cells remained in contact with it continuously from the beginning of the experiment on the living cells right through until reactivation. The most probable explanation is that the inactivation of the cilia is due to blocking of metabolic reactions leading to ATP synthesis, and that it occurs when the reserves of high-energy compounds in the cell are used up.

Aleksandrov and Lieh Suan-mai (1968) placed models of ciliated cells of the frog palate in solution containing 1×10^{-3} M 2,4-dinitrophenol, which abolishes ciliary motility in living cells. However, this solution did not prevent ciliary function in models on the addition of ATP. This experiment indicates that dinitrophenol, in a concentration sufficient to inhibit ciliary motility in living cells, does not disturb their contractile apparatus.

As long ago as in 1954 Hoffmann-Berling investigated the action of monoiodoacetate on anaphase models of fibroblasts. In a concentration of 1×10^{-2} M monoiodoacetate suppresses anaphase movement of chromosomes in living cells. However, this inhibitor in the same concentration does not prevent anaphase separation of the chromosomes in model cells. Consequently, it does not act on the mitotic motile apparatus. Perhaps it inhibits mitotic movement in living anaphase cells because it blocks ATP synthesis.

Other well-known inhibitors of mitosis such as 5×10^{-4} M colchicine, 1×10^{-3} M aminopterin, 5×10^{-3} M caffeine, and 1×10^{-1} M urethane likewise do not act on the anaphase model of fibroblasts. It is clear that in the living cell these poisons also do not act on the motile elements of the spindle, but on certain other systems which are not present in models but which are essentially components of the mechanism of mitosis in the living cell (Hoffmann-Berling, 1954b).

Models have been used with success to analyze changes in the resistance of cells to the action of injurious factors. A number of agents of widely different character, ranging from glycerol through hypertonic saline solutions to heavy water (D_2O), substantially modify the resistance of the ciliary function of frog ciliated epithelial cells to the injurious action of heating and of a high hydrostatic pressure. Under the influence of some of these

agents, resistance is increased to both these factors. In some
cases (hypertonic saline solutions) resistance to pressure is increased but resistance to heat is sharply reduced. Sometimes
the direction of the changes depends on the period of action
of the agent affecting resistance. To investigate whether the
change in resistance of the cilia of living cells is connected with
stabilization of their contractile apparatus, model experiments
have been used (Arronet, 1964a). Attempts were made to discover
whether the ability of cells with experimentally increased resistance to the corresponding injurious agent to yield models is itself
increased. In other words, is it possible to obtain working models
from such stabilized cells after exposure to a higher intensity than
if the agent is applied to ordinary cells? Similar experiments
were carried out to study the action of agents which reduce the
resistance of ciliated cells to injury. The results showed that in
all cases investigated changes in the resistance of the "ability of
the cells to yield models" were absolutely identical with the changes
in resistance of ciliary motility in living cells, as Tables 4 and 5
clearly show. The facts given in Table 5 are particularly demonstrative. The character of the changes in thermostability of ciliary
motility under the influence of 2 M glycerol in Ringer's solution.
depends on the time of incubation of the cells in this solution. If
contact of the cells with the glycerol solution is brief, and the glycerol solution is brief, and the glycerol probably acts simply as a

TABLE 4. Stabilization of Cells of the Palatial
Ciliated Epithelium of Rana temporaria and of
Their Contractile Apparatus Relative to the
Injurious Action of a High Hydrostatic Pressure

Stabilizing agent	Resistance to pressure (in atm)	
	of living cells	of ability to yield models
Without stabilization (control)	1550—1650	1560
2 M glycerol during exposure to pressure (5 min)	2800	2575
Ringer's solution of 8 times normal strength (10 min before exposure + 5 min during exposure)	2220	2250

TABLE 5. Changes in Resistance of Cells of the
Palatal Ciliated Epithelium of Rana temporaria
and of Their Contractile Apparatus to Heating
for 5 min Produced by Exposure of the Cells
to 2 M Glycerol for Different Periods

Duration of contact of cells with 2 M glycerol	Thermostability (in °C)	
	of living cells	of ability of cells to yield models
No contact with glycerol (control)	41.9	40.9
During exposure to heat only	37.8 ± 0.2	37.1 ± 0.2
4 h before exposure + 5 min during exposure to heat	42.4 ± 0.07	42.4 ± 0.08
21 h before exposure + 5 min during exposure to heat	43.0 ± 0.08	42.5 ± 0.1

nonspecific hypertonic agent, the thermostability of the cilia falls. Later, if the cells remain longer in glycerol solution, their thermostability rises; this is evidently due to penetration of glycerol into the cells. Similar bidirectional changes also take place in the thermostability of the ability of the cell to yield models: to begin with a sharp decrease, and a subsequent increase. It could thus be concluded that changes in the thermostability of the contractile apparatus of the cilia, at least in the case under discussion, lie at the basis of those changes in thermostability of the cells which are judged from the suppression of ciliary motility.

In some cases when it was shown that the resistance of the cells to injury could be increased, it was not demonstrated whether the resistance to model formation also was increased. However, agents increasing the resistance of living cells (as judged by the preservation of their ciliary motility) were shown to increase the resistance of models provided the models were made previously, subjected to the action of the stabilizer, and then while the stabilizer continued to act, exposed to injury. For instance, resistance of both cells and their models to heating was obtained by the action of D_2O (Aleksandrov, Arronet, Den'ko, and Konstantinova, 1965, 1966).

Models of ciliated epithelium, like cells, are stabilized relative to injury by hydrostatic pressure by treatment with glycerol. However, the models have no permeability barriers, and the phase of hypertonic action of glycerol is not therefore reflected in the changes in their resistance, so that specific stabilization due to penetration of glycerol into the models appears immediately.

In all probability there is a causal connection between changes in the resistance of the contractile apparatus and changes in the resistance of ciliary function to injury, i.e., the former lie at the basis of the latter. This conclusion follows from the facts described above.

An example of the use of muscle models to analyze the increase in resistance of muscles obtained experimentally will now be given. The viability and stability of a muscle relative to various denaturing agents can be increased if the muscle is moderately stretched. This phenomenon was investigated in detail by Ganelina, who used glycerinated muscle models to determine whether stretching affects the resistance of the system of contractile proteins of the muscle. She found (Ganelina, 1962) that models also have increased resistance to heat if they are exposed to a raised temperature in a stretched state. Models which were stretched during preparation possessed greater contractility and higher ATPase activity. Ganelina considers that the increased resistance of stretched muscles which she found is due to stabilization of their proteins. This conclusion was expressed in more concrete terms later by Vorob'ev and Ganelina (1963), who postulated that moderate stretching stabilizes the orderly configuration of the fibrillary muscle proteins, with the result that the stability of the proteins and of the muscles themselves is increased.

Contractile models have also been used in cytoecology. It was stated above that the degree of thermostability of ciliated epithelial cells is a distinctly species-specific feature. This has also been found to be true of other cells and tissues and, in particular, of muscle tissue. Results obtained on a wide range of botanical and zooligical material have led to the conclusion that in closely related species which differ in their temperature ecology (meaning the temperature of their habitat for poikilothermic animals and their body temperature for homoiothermic animals) the thermostability of analogous cells correlates with the degree of thermophilia of the species (Aleksandrov, 1952, 1963, 1964, 1967, 1969; Ushakov, 1956, 1967).

Material relevant to this issue will be found in Table 2 (see page 119). This gives temperatures reflecting the resistance of ciliated epithelial cells of 3 species of frogs to heating for 5 min. These frogs differ in their territory of distribution. The habitat of Rana ridibunda is in the south of the USSR; Rana camerani also lives in the sourth, but in the mountains, i.e., in a colder climate; Rana temporaria lives farther north than the other two species, and its habitat even extends beyond the Arctic Circle. To correspond with these differences of habitat, the 3 species of frogs differ in their thermophilia, by which they are arranged in the following order: Rana ridibunda > Rana camerani > Rana temporaria. It follows from Table 2 that the frogs of these species are also arranged in the same order as regards the level of thermostability of the ciliated cells of their palatal and oviductal epithelium. This is just one example to illustrate the above rule regarding the correlation between thermostability of the cells and thermophilia of the species.

The same table also shows that the frogs of these species also lie in exactly the same order in the thermostability of their contractile apparatus, as judged from the possibility of obtaining working models from cells heated to different levels. The information given in Table 2 could be supplemented by an additional column containing details of the resistance of models previously obtained from cells to heating for 5 min. These figures for models of palatal cells are as follows:

   Rana ridibunda . . . . 42.8°C
   Rana camerani . . . . 41.4°C
   Rana temporaria . . . . 38.6°C

The corresponding figures for models of ciliated oviductal cells are:

   Rana ridibunda . . . . 41.6°C
   Rana camerani . . . . 39.5°C
   Rana temporaria . . . . 38.2°C

It will easily be seen that in the thermostability of models of their ciliated cells the frogs of these species lie in precisely the same order as their thermophilia. Since contractile models can be regarded, albeit with certain reservations, as preparations of contractile proteins, the existence of correlation between the level of thermostability of ciliated cell models from animals of a given species and the thermophilia of the species provided a powerful argument in support of the cytoecological conclusion that the environmental temperature of the species corresponds not only to the thermostability of its cells, but also to the thermostability of its intracellular proteins.

Facts relating to the thermostability of various protein preparations, including glycerinated models of ciliated cells and muscles, have become a regular feature of cytoecological papers published in recent years. As a rule this has been dictated by the investigator's endeavor to discover whether particular changes or differences in the thermostability of the cells are accompanied by corresponding changes or differences in the thermostability of the cytoplasmic porteins.

As an example, in addition to several regular rises and falls of thermostability of the muscles of juvenile frogs during the course of the year, Chernokozheva (1967b) discovered seasonal changes of the same character in the thermostability of models obtained from these muscles. She emphasizes that correlation exists between the changes in thermostability of the muscles and of their models. However, the models are more stable than the muscles, so that the thermostability of the models in no way limits or determines the thermostability of the muscles. It can only be assumed that parallel seasonal changes take place in the thermostability of the muscles (as defined by thermal inhibition of their ability to respond by contraction to electrical stimulation) and of their contractile apparatus.

Conversely, changes in the thermostability of frog muscles connected with the age of the animals are unaccompanied by any changes in the thermostability of models of these muscles (Chernokozheva, 1967a). There is likewise no correlation between the individual changes in thermostability of muscle tissue and of model muscle fibers. This has been shown by Ushakov and Pashkova (1967) on the mantle muscle of the Black Sea mussel.

Dregol'skaya and Chernokozheva (1969) discovered that seasonal changes in the thermostability of ciliated epithelial cells of

Rana temporaria correlate with changes in the thermostability of models obtained from these cells. Similar correlation was found by these workers as regards seasonal changes in thermostability of frog muscles and models of their fibers.

Makhlin and Skholl' (1968) showed that if Barents Sea mussels and the bivalve mollusk Macoma are kept at 18°C, which is higher than their usual temperature, this leads to an increase in the thermostability both of their muscles (stability was tested by thermal inhibition of the ability of the muscles to contract in response to electrical stimulation) and of their contractile apparatus (tested by inhibition of the contractility of models). However, keeping the mollusks at a low temperature (+2°C) either causes no change in the thermostability of the muscles (mussel) or actually reduces it (Macoma), although at the same time it increases the thermostability of their models.

By using a sib-selection technique, Skholl' (1970) selected voles by the character of the high thermostability of their muscle tissue. After several generations she obtained animals whose muscles possessed increased thermostability by comparison with those of voles not artificially selected. She found that the thermostability of models obtained from the same muscles of these voles also was increased. In this case, therefore, correlation was found in the individual variation in thermostability of the muscles and of their contractile apparatus.

There is yet another way in which models can be used (perhaps it would be better to say must be used), although at present this is not being done on a sufficiently wide scale. This is the use of contractile models for teaching purposes, for practical exercises in courses of cytology or, perhaps, biochemistry. In an early publication on model muscle fibers (Szent-Györgyi, 1951) it is strongly recommended that these models be demonstrated to students. Szent-Györgyi is particularly concerned with how such exercises should be conducted, and how contraction of the glycerinated fiber should be induced using muscle juice instead of ATP, because at that time ATP was not yet obtainable commercially. Nowadays it is much easier to organize a practical course and to give demonstrations using models because ATP is widely available.

Besides glycerinated muscle fibers, glycerinated models of frog ciliated epithelium can also be recommended for teaching purposes. The technique of their preparation and observation is extremely simple, the reagents and materials required for microscopic examination are widely available, and these models can readily be obtained. The reactivation of these cell fragments, dead, which appear so completely with a drop of ATP leaves a striking impression. It is difficult to image anything so comparable in its vividness and clarity as this demonstration of the role of ATP not only in mechanochemical processes, but in general as a source of accumulated energy.

The first occasion on which Aleksandrov and the present writer observed the "reanimation" of ciliated epithelial models by the action of ATP in 1956 made a tremendous impression on both of us. Whenever we have demonstrated these models to all sorts of people we have always noticed the admirable impression made on the audience by this effective scientific fact.

Together with experiments to reactivate ciliated epithelial cells, practical exercises to show the action of metabolic inhibitors such as dinitrophenol, monoiodoacetate, and azide on cell motility should be carried out. Bearing in mind what was written above (page 123) two types of experiments should be performed in order to show that: a) dinitrophenol or another of the inhibitors stops ciliary movement of living cells of the frog ciliated epithelium, and b) that the same inhibitor, used in the same concentration, does not prevent ciliary movement in models of ciliated cells induced by ATP together with Mg^{++}.

Concrete details of practical exercises for teaching purposes in line with these recommendations are given in Chapter IV.

Chapter IV

Methods of Obtaining Motile Models

The general principles of obtaining models and the role of the individual components of the solutions used are described in Chapter I. Instructions for the preparation of models of particular objects are given below.

PRELIMINARY REMARKS

Abbreviations used:

AM — actomyosin.
ATP — adenosine triphosphate, the tetrasodium salt of adenosinetriphosphoric acid.
EDTA — sodium ethylenediamine tetra-acetate (versene, Trilon B).
EGTA — sodium ethylene-glycol tetra-acetate.
ES — extracting solution.
RS — reactivating solution (inducing movement).
WS — washing solution.

The glycerol concentration is given everywhere in vol.%. The vague expression "in the cold" used by the authors and reproduced by the present writer in the same indefinite form should be understood as meaning "at a temperature close to 0°C, which need not be maintained too strictly."

Where the composition of the WS is not shown, it can be based on the composition of the ES, leaving out glycerol and/or lytic agents, or on the composition of the RS, leaving out ATP and $CaCl_2$.

Now that Tris-maleate, Tris-HCl, thioglycollate, and other new buffer mixtures are available, in many cases the phosphate buffer which precipitates Mg^{++} and Ca^{++} can profitably be replaced by these other buffer solutions. However, it must be remembered if this is done that phosphate can also play the role of plasticizer of the models, and this is often an essential factor determining successful function of the model.

ACTOMYOSIN THREADS

Actomyosin Threads from Cross-Striated Skeletal Muscles

Obtaining the actomyosin (Portzehl et al., 1950). All procedures are undertaken at a temperature of around 0°C. After careful mincing and grinding of the femoral muscles or psoas muscle of a rabbit (using a mincer and various types of homogenizers) the mince is extracted with 3 volumes of 0.6 M KCl at pH 6.8-7.0 for 24 h, using a mixer. This is followed by centrifugation, decantation, and precipitation and reprecipitation of the actomyosin. By superprecipitation the protein concentration can be increased to 1.5-2.7%.

Preparation of the threads. A glass cannula with capillary tube, having a lumen 100-200 μ in diameter,* is fixed to an all-glass (Luer) syringe. AM is expelled from the syringe through this capillary tube into CO_2-free distilled water cooled to 2°C, and poured into a rotating glass dish. Threads are formed.

For nonquantitative experiments, such as demonstration simply that the thread can contract, these procedures are all that is necessary. The thread is transferred to a drop of solution consisting of 5×10^{-2} M KCl + 1×10^{-3} M $MgCl_2$. On the addition of 0.05 ml of 1% ATP solution to this drop, the thread will contract.

For investigations requiring measurement of tension (contraction) developed by the thread in response to the action of ATP

---

*A cylindrical capillary tube without a tapering end probably ensures a laminar flow, so that the protein molecules are oriented longitudinally; the degree of their orientation must also increase with an increase in length of the capillary tube.

under isometric conditions and for other quantitative investigations it is necessary to prepare threads with a higher degree of orientation of the AM molecules, with an increased concentration of AM and, consequently, with a higher level of intermolecular interaction, greater breaking strength, and increased strength of contraction. For this purpose the threads obtained as described above must be treated by one of the following methods.

By the method of Portzehl (1951). The threads are kept in water at 0°C for 20 min (or longer) and then transferred to 30% glycerol solution in 0.01 M phosphate buffer, pH 6.7-7.0, for 30 min. The glycerol solution is made up from bidistilled absolute glycerol. The threads are then attached to a strain gage (dynamodilatometer) and are stretched overnight in a moist chamber at ~3°C; under these conditions they dry out slowly and uniformly. A check on their birefringence and thickness is carried out with a polarization microscope. Their length is measured and must be more than twice its initial value. The diameter must be reduced to 70-80 μ. The protein concentration in such threads is 5-10%. Without taking the threads from the lever of the strain gage they are transferred to glycerol-free WS, in which they remain for a few minutes, after which the experiment with RS can begin. The optimal ATP concentration in RS for the function of these threads is $(2-3) \times 10^{-3} M$.

By the method of Hayashi (1952). Langmuir dishes 8 cm long, 10 cm wide, and 1.6 cm deep are used. The edges are covered with a uniform layer of ceresine (purified, decolorized paraffin wax). The dish is filled to the brim with 0.05 M KCl in veronal buffer, pH 7.0. Copper strips (glass can also be used) measuring $25 \times 0.6 \times 0.6$ cm are placed on top to act as movable barriers. The AM solution containing 50% glycerol is diluted 2-3 times (depending on the AM concentration) with 1 M KCl. Drops of AM solution are applied to the phase boundary between 0.05 M KCl and air between the barriers. In this way a stretched surface film is formed. After waiting for 5 min until the protein has spread to form a monomolecular layer the film is removed by slow (so that air bubbles are not trapped in the film) movement of the barriers. The film, thus compressed into a thread, is cut free from the barriers. The thread floats freely on the surface of the liquid and is ready for work.

Actomyosin Threads from Smooth Muscles

(After Dörr and Portzehl, 1954)

<u>Obtaining the actomyosin.</u> All procedures are carried out at a temperature of around 0°C. The yellow adductor muscle of Anodonta is minced in a Waring blender for 4 min. The ionic strength of the suspension is adjusted to 0.6 with KCl and the pH to 7 with 0.01 M phosphate buffer, after which the suspension is treated with 5×10^{-3} M KCN and left for 12 h at 3°C. The extract is centrifuged at 22,000 g until it is transparent. Its AM concentration is then ~1.2% and the sensitivity of its viscosity to the action of ATP, determined by the formula

$$\frac{\log \eta - \log \eta_{ATP}}{\log \eta_{ATP}} \cdot 100,$$

is 100%.

The AM is reprecipitated twice by tenfold dilution with water and subsequent centrifugation. Unlike AM of rabbit skeletal muscles, the first precipitation in the absence of ATP occurs only after 15-20 min, and in the presence of ATP later still. The residue of the second precipitation is superprecipitated with 1×10^{-4} M ATP and concentrated by centrifugation. It is then dissolved by the addition of 1×10^{-2} M ATP in a volume of between 1/4 and 1/5 of the residue and of crystalline KCl in an amount providing an ionic strength $\mu = 0.6$. The solution contains 2-2.5% protein, and becomes gelatinous after hydrolysis of the ATP.

<u>Preparation of the threads.</u> AM is extruded through a capillary tube into a cold solution of 8×10^{-4} M $MgCl_2$ and KCl in a concentration giving $\mu = 0.06$, pH 6.9. The threads thus formed are allowed to stand in this solution for 20 min. The protein concentration in the threads must be increased to about 5% by soaking them with 30% aqueous glycerol and drying them in a moist chamber at 4°C. The threads are stored in 80% glycerol buffered to pH 6.8-7.0. Immediately before the experiment the threads are stretched to twice their length on a strain gage in a solution with $\mu = 0.06$ and pH 6.9, checking their tension and birefringence as this is done. The stretching increases the degree of longitudinal orientation of the filaments in the threads. The diameter of the thread must be less than 60 μ.

Paramyosin (A-tropomyosin, TM_A) threads from mollusk smooth muscle (Rüegg, 1961b). The smooth-muscle parts of the adductor muscle of Pecten maximus or the retractor muscle of Mytilus edulis are used. Crystalline tropomyosin A is obtained as described by Bailey (1956). To obtain Hayashi's monolayer threads a drop of 0.1% TM_A solution in 0.5 M KCl is allowed to spread over the surface of 40% saturated $(NH_4)_2SO_4$, pH 6.5 is created by NH_3. With two glass rods sliding on a Langmuir dish the monolayer is compressed into a filament. About 10 filaments are collected into a "cable," which is dried in the air, washed with 50% ethanol and 50 mM KCl, and then dried again under slight tension and stored in 80% glycerol at $-9°C$ until required for the experiment.

To obtain threads by the urea method, crystals of TM_A prepared by centrifugation at 15,000 g are dissolved by the addition of an equal volume of 10 M urea and the pH is adjusted to 7 by histidine buffer. The viscous solution is extruded under pressure from a syringe through a capillary tube into 30 mM phosphate buffer, pH 6. In this way brittle threads are formed but they become elastic in a solution of 50 mM KCl, histidine buffer, pH 7.0. They are air-dried under slight tension and stored in 80% glycerol in a deep-freezer.

When working with the threads, various designs of strain gages are used, the same as in experiments on model fibers. The only difference is that when working with threads the lever with the pen must be smaller and lighter because of the smaller tension developed by the thread compared with the fiber. The thread must also be attached not with clips, as in the case of the fiber, but by varnish or inert glue. We are familiar with the strain gages of Weber (1951), Hayashi (1952), Ranney (1954), Efimov (Lyudkovskaya and Kalamkarova, 1965), and Chaplain (1966). Pringle (1967) has given a detailed account of his strain gage (see also Jewell and Rüegg, 1966).

MODELS OF THE MUSCLE FIBER

Models of Cross-Striated Skeletal Muscle Fibers

(After Szent-Györgyi, 1949, 1951.) The most convenient object is the rabbit psoas muscle. This muscle consists of long, parallel fibers which are easily separated. The gracilis and sartorius muscles of the frog can also be used.

The muscle of a freshly killed rabbit is split up with blunt forceps into bundles about 1 mm in diameter. A thin rod is placed alongside the bundles which are secured to it at their ends by silk threads, thus indicating their normal length. The bundles are then cut from the muscle and immersed together with the rod in 50% glycerol solution at 0°C. After 24 h the bundles together with the ligatures are cut away from the rod and placed in 50% glycerol solution at 0°C for a further 24 h; for prolonged storage the bundles are kept in 50% glycerol at −20°C.

The bundles are split up into thinner groups of fibers (or into single fibers) by means of watchmaker's forceps after the bundles have been kept for about 1 h in 15-20% glycerol.

Fibers or groups of fibers contract when placed in 0.2% ATP. To demonstrate actual contraction of the muscle fiber, muscle juice can be used instead of ATP. To obtain it the muscle (for example, the remains of the same psoas muscle) are minced and the mince is mixed with an equal volume of boiled water. The suspension is rapidly heated to 80-100°C. The mince is pressed through cloth and the juice is ready for use. This method is very convenient and vivid when used for practical demonstrations.

(After A. Weber, 1951.) The psoas muscle of a freshly killed rabbit aged 8-12 months is stimulated by an induction current until the muscle is completely fatigued (this prevents contracture of the muscle fibers when placed in extracting solution).[*] The connective-tissue membrane is detached and bundles of fibers 1-2 mm thick are prepared. Injury to the fibers must be avoided. The fibers are tested for absence of injury by determination of the uniformity of their birefringence under the polarization microscope.

The bundles are placed for 12-24 h in 50-100 volumes of a solution of 30% glycerol in 0.01 M phosphate buffer, pH 7.0, at 0°C. They are then transferred into a solution containing 96% glycerol in 0.01 M phosphate buffer, pH 7.1, in which they are kept for 1-3

---
[*]Instead the bundles of fibers or, in the case of frog muscles, the whole muscles can be tied to stretchers so that isometric conditions are created and the muscle fixed in its initial length. The ileo-fibularis and sartorius muscles of frogs have also been used by the writer and are convenient for preparing models of single fibers.

weeks at 0°C. Single fibers are dissected from the bundles under a binocular loupe in a solution containing 50% glycerol in 0.01 M phosphate buffer, pH 7.0, after which they are kept for at least 12h in a solution of 96% glycerol. The fibers are then washed free from glycerol with the WS: 0.01 M phosphate buffer, pH 6.9, and 1×10^{-3} M MgCl$_2$. Contraction is induced by ATP, diluted before the experiment with WS; the final ATP concentration is of the order of 4×10^{-2} M.

Preparation of Model Fibers of the Flying Muscles of Insects

(After Abbott and Chaplain, 1966)

This method must supersede many of those at present used for the preparation of a wide range of motile models because it ensures the virtually complete removal of noncontractile proteins and of contractile nonmyofibrillary membrane proteins from the objects. It may perhaps enable models to be obtained from new objects.

The material consists of the dorsal longitudinal and oblique muscles of giant tropical aquatic beetles belonging to the Belostomidae family: Lethocerus cordofanus, L. maximus, Hydrocyrius columbiae, etc. The insect is placed whole in ES-1: 50% glycerol in 67 mM K-phosphate buffer, pH 7.0. The insect is infiltrated with ES-1 at reduced pressure so that the air in the tracheoles is replaced by the ES. This is perhaps easiest done by placing the animal and the ES in a large syringe in which a negative pressure is created by withdrawing the closely ground plunger while the orifice is hermetically closed (Aleksandrov, 1954). The insect is kept in ES-1 for 24 h with agitation, after which the solution is changed. After a further change of solution 24 h later the preparation can be kept at -18°C for 1-8 months. Bundles of 15-25 fibers are cut from the glycerinated preparation so that fibers remain attached to each other at one end but are separated throughout their length as far as the other end. The bundles are then placed in one of two ES-2 solutions of the composition given below for 18 h at 0-2°C. ES-2 (1): 0.5-0.6% Tween-80, 5 mM MgCl$_2$, 47 mM K-phosphate buffer, pH 7.5-7.8. ES-2 (2): 0.5-0.6% Tween-80, 0.88 M sucrose, 1 mM MgCl$_2$, 10 mM Tris-HCl buffer, pH 7.1-7.2. They are again transferred to the ES-1. The model fibers are ready for use. They can be used as they are or models of isolated myofibrils can be prepared from them. This will be described here for convenience, although the subject of model myofibrils is examined later (see page 141).

In the latter case the model fibers are homogenized with a Teflon pestle (at low speed) in a solution of 0.34 M sucrose, 1 mM EGTA, 15 mM Tris-HCl buffer, pH 7.2. Centrifugation follows at 600 g for 20 min. The residue is washed with the same solution and recentrifuged at 600 g. The residue contains model myofibrils. These only require washing with 0.12 M KCl and centrifugation for 15 min at 800 g. All procedures must be carried out at 0-4°C.

The following solutions are used in experiments with models of fibers and myofibrils. Solution A inducing contracture: 5 mM MgCl$_2$, 4 mM EGTA, 70 mM KCl, 20 mM Tris-HCl. Solution B inducing relaxation: 5 mM ATP, 5 mM MgCl$_2$, 4 mM EGTA, 60 mM KCl, 20 mM Tris-HCl. Solution C inducing oscillation of the models: 5 mM ATP, 5 mM MgCl$_2$, 4 mM Ca-EGTA, 60 mM KCl, 20 mM Tris-HCl, pH 7.15 (as also in solutions A and B). By mixing solutions B and C in different proportions, solutions with different concentrations of Ca^{++} as required can be obtained.

Models of Smooth-Muscle Fibers

<u>From the white and yellow adductors of Anodonta</u> (Ulbrecht and Ulbrecht, 1952). The adductor of the unanesthetized mollusk is detached together with the shell. The muscle, hanging between two pieces of shell, is split into thin bundles each about 1.5-2.0 mm in diameter.

<u>From the circular layer of the muscle wall of the forg's stomach</u> (after Ulbrecht and Ulbrecht, 1952). The stomach is excized and a glass rod thick enough to prevent contraction of the stomach during dissection of its circular layer is introduced into it through the stump of the esophagus.

<u>From the cow's rectum and from various retractors</u> (after Ulbrecht and Ulbrecht, 1952). The muscle is cut into strips which are stretched to their resting length. Extraction is carried out in this state.

The single model fibers are extracted and dissected in the same way as during preparation of models of cross-striated muscles by Annemarie Weber's method, but at $-9°C$, and because of this the glycerol concentration in ES is raised to 45%. Cysteine (2×10^{-2} M) is added to the ES. The extraction time must not exceed 14 h. If the models must subsequently be stored, this can be done in concentrated (d = 1.23, equivalent to 89%) glycerol for several weeks. The WS and RS are the same as for models of cross-striated muscles, but with the addition of 2×10^{-2} M cysteine. Usually the RS contains 2.5×10^{-2} M ATP.

Models of smooth-muscle fibers of blood vessels (after Bohr et al., 1969). Arterioles about 200 μ in diameter with longitudinal fibers are used, but the helically climbing fibers of the pig's carotid artery are particularly convenient. The media at the severed end of the vessel is grasped with forceps and stripped from the vessel. A narrow strip of fibers climbs helically in the wall of the vessel to form its muscular layer. On stretching the helix uncoils, and this provides the investigator with a thin bundle of fibers requiring no further dissection. Segments 5-10 mm in length are secured by their ends to glass rods so that the fibers cannot contract and extraction is carried out with 50% glycerol at 0°C for 48 h, after which the fibers can be kept in more concentrated glycerol at $-18°C$. After extraction the model fibers are stretched under a load of 200 mg for 12 min in WS consisting of: 50 mM KCl, 5 mM MgCl$_2$, 0.1 M Tris buffer, pH 6.5. Reactivation in RS consisting of: 6-20 mM ATP, 50 mM KCl, 5 mM MgCl$_2$, 0.1M Tris buffer, pH 6.5. In RS containing 6 mM ATP the model develops a mean tension of 30 mg-wt.

Models of intestinal smooth-muscle fibers (after Beck et al., 1969). The small intestine of young (under 2 cm in length) frogs (Rana temporaria), 600-900 μ in diameter, is extracted with the solutions used by Kamiya and Kuroda when preparing models of myxomycete plasmodia (see page 162). The glycerinated intestine is then cut up into sections 1-2 mm long washed to remove glycerol, and placed in RS: 5 mM ATP, 5 mM KCl, 30 mM NaCl, 10 mM Tris buffer, pH 7.0. After a stay of 15 min in the RS the model fibers of the circular muscles contract to such an extent that the diameter of the intestinal lumen is reduced by 42-63% (i.e., to 58-37%) of its initial value. Models of longitudinal fibers, also contained in the preparation, contract only very slightly, so that the segment of glycerinated intestine is shortened by only 5-10% (i.e., to 95-90%) of its initial length or simply bends.

Electron-microscopic and phase-contrast investigations of this model revealed the same protofibrils as in preparations made from the living intestine, but they are less orderly in their orientation than those in invertebrate smooth muscles.

This easily made model could perhaps be improved by the use of double extraction as described by Abbott and Chaplain, i.e., by glycerol extraction and subsequent detergent treatment (see page 137).

Model fibers of smooth obturator muscles of lamellibranch mollusks (Rüegg, 1961b). The smooth-muscle parts of the adductor muscle of Pecten maximus or the retractor muscle of Mytilus edulis are used. The valves are opened by mechanical stimulation to the region of the mouth, and tonic contraction of the adductor or retractor is prevented by rapid cutting of the corresponding nerve. The smooth part is carefully separated from the striped, tetanic part (from which models of fast fibers can be prepared at the same time) and from the soft tissues of the valves, but it is left attached to the shell and placed in a solution of the following composition: 1 volume sea water + 1 volume isotonic $MgCl_2$. Total relaxation of the muscle takes place. It is then transferred to the ES: 1 volume glycerol + 1 volume of aqueous solution of 50 mM KCl and 20 mM histidine buffer, pH 7. Bundles each 2 mm in thickness, are dissected under a layer of this solution and transferred to the same ES, but cooled to $-9°C$. Extraction at this temperature continues for several days, after which bundles 0.1 mm in thickness are dissected. These can be reactivated by means of RS containing ATP, during which the tension, shortening, and thickness of the model are measured. These models will hydrolyze ATP at the rate of $(5-6) \times 10^{-2}$ μmole ATP/min per gram protein or 1×10^{-7} mole ATP/min per cubic centimeter of model.

Preparation of Glycerinated Models of Muscle Fibers from the Ventricular Myocardium

(After Lee, 1961.) This worker describes methods of dissection and extraction, gives drawings of a chamber where reactivation can be observed, and gives results showing the increase in tension developed by these models under the influence of ouabain. Such models still contain soluble relaxing factor. Possibly these models also would be better prepared by extraction by the method of Abbot and Chaplain (see above, page 137).

Strain gages for studying the work of model fibers are discussed above (see page 135).

ISOLATED MYOFIBRILS

By the method of Schick and Hass (1949). The material consists of skeletal and heart muscles of rabbits, guinea pigs, dogs, and man.

A block measuring $25 \times 25 \times 10$ mm is excised from the proximal segment of a rabbit's forelimb, frozen rapidly, and cut into sections 15 μ thick on a freezing microtome. The sections are placed in 250 ml potassium phosphate-citric acid buffer solution (pH 6.4, $\mu = 0.25$), cooled to 0°C, containing 5 ml of a 0.4% filtered solution of commercial trypsin. The sections are periodically taken from the solution and examined under the microscope. When slight pressure on the cover slip is sufficient to break the sarcolemma and liberate the myofibrils (usually after digestion for 30-45 min) the sections are filtered off, washed, resuspended in phosphate-citrate buffer (pH 6.4, $\mu = 0.25$), and homogenized for 5-10 sec in a Waring blender. The sarcolemma is ruptured and the myofibrils liberated. The preparation is again washed with the same solution and centrifuged at 0°C. The final suspension is a white floccular mass containing only traces of collagen and fragments of fibers as impurities visible when the mass is examined under the microscope.

When the isolated myofibrils are placed in 0.075% ATP solution at room temperature they contract rapidly and are converted into small, round bodies which can be observed under the microscope.

By the method of Perry (1951). The material consists of the psoas muscle of a rabbit. From 2 to 3 g muscle tissue is excised, the block is kept for 30 min at room temperature in Ringer's solution, and sections 25 μ thick are cut on a freezing microtome. The sections while still frozen are dipped in 0.08 M borate buffer, pH 7.1, at a temperature just above freezing point and 25 mg purified collagenase is added. Digestion of the sarcolemma takes place at 0°C for 3-7 h (the time depends on the quality of the collagenase). The resulting suspension is diluted with 150 ml borate buffer and homogenized in a Waring blender for 20 sec. All the fibers must be disrupted to myofibrils and this is verified by examining a drop of suspension to which a drop of methylene blue is added under the microscope. The myofibrils are washed 3 times with borate buffer and then centrifuged at 1500 g. The second and third supernatants are transparent. The residue is resuspended in 100 ml borate buffer and centrifuged for 1 min at 400 g. The residue is discarded and the suspension again centrifuged at 1500 g, and the residue obtained is again resuspended in 100 ml buffer and centrifuged for 1 min at 400 g. All procedures are carried out at 0°C. The resulting cloudy supernatant is a slowly sedimenting suspension of myofibrils containing usually 1-2 mg protein/ml. Most myofibrils are single but some are in bundles. Reactivation is accomplished with $4 \times 10^{-4}\ M$ ATP and $2 \times 10^{-3}\ M$ $MgCl_2$.

This method is preferable to that using tryptic digestion because trypsin, unlike collagenase, injures myosin and interferes with its ability to combine with actin.

By the method of Portzehl (1954). Material: rabbit psoas muscle. The muscle is left in situ until rigor mortis has developed when it is excised, cooled to 0°C, and homogenized in a Waring blender with 10 volumes distilled water for 2.5 min. The residue is separated by centrifugation, then suspended in 10 volumes of $0.1\ M$ KCl, recentrifuged and again washed in $0.1\ M$ KCl, followed by a further centrifugation. Only the upper layer of the last sediment, which consists of myofibrils, is taken. This material is placed in concentrated glycerol (d = 1.23) in which it is stored.

Before the experiment the suspension is diluted 45 times with water. (Probably KCl should be added to give a solution with ionic strength $\mu = 0.12$). Contraction is induced with a solution of $5 \times 10^{-3}\ M$ ATP, $5 \times 10^{-3}\ M$ $MgCl_2$, $1 \times 10^{-2}\ M$ phosphate buffer, pH ~ 7, and KCl in a concentration giving an ionic strength of the order of 0.12.

The degree of shortening of the myofibrils is estimated by measuring the length of 100 myofibrils before and after contraction and comparing the means. The phase-contrast microscope is used. To determine the rate of contraction, it is recommended that motion pictures be taken and the changes in length of the myofibrils measured on the frames.

By the method of Hasselbach (1952). Material: frog femoral muscles and rabbit psoas muscle. The muscles are taken from a recently killed animal and homogenized in the cold in a Waring blender in 10 volumes of 30% aqueous glycerol. The total time of homogenization is 10 min, but it is split up into 5 periods each of 2 min so as to avoid heating by friction; in the intervals between the 2-min periods of homogenization the vessel containing the material is placed for 15 sec in a refrigerator at a low temperature. The mince is allowed to stand for 24 h in the cold for extraction. It is then washed repeatedly with CO_2-free water (~3°C) and centrifuged, yielding a suspension of myofibrils. (Probably only the upper layer of the residue should be suspended after the last centrifugation. Hasselbach does not state to what extent the myofibrils are preserved in his preparation).*

Isolated myofibrils from mollusk smooth muscles (Rüegg, 1961b). The same smooth-muscle parts as in the preceding paragraph are used. The muscles are minced in a 30-ml blender in a solution of: $0.15 M$ KCl, 0.4 sucrose, $0.02 M$ histidine buffer, pH 7. Centrifugation. The fraction between 1000 and 2000 g consists almost entirely of myofibrils.

For details of the preparation of myofibril models from the flying muscles of insects, see page 137.

THE MUSCLE FIBER WITHOUT ITS SARCOLEMMA

(After Costantin and Podolsky, 1967.) A bundle of 6-10 fibers is taken from the semitendinosus muscle of a frog, gently wiped dry with filter paper, and covered with mineral oil on a glass slide. One of the fibers (the thickest), 80-100 μ in diameter, is incised and, the sarcolemma removed over a small area by hand. The layer of superficial myofibrils is also stripped off with the sarcolemma. The fiber is uncovered for a distance of 1 mm, and if the sarcolemma is wound into a roll as this is done, the distance can be even greater. The diameter of the fiber after removal of the sarcolemma is reduced by only 10%. For the experiment the fiber is then placed under a microscope (objective 10× or 40×). The preparation is suitable for use for experimental purposes only for 40-60 min after re-

*See Notes Added in Proof, page 172.

moval of the sarcolemma, for in the exposed area a contracture develops. On the appearance of the first signs of this contracture the experiment must be abandoned.

MUSCLE FIBER WITH DISRUPTED COUPLING IN THE SARCO-
TUBULAR SYSTEM

(After Eisenberg and Eisenberg, 1968.) Sartorius muscles of frogs, attached to a stretcher, are immersed in Ringer's solution containing 0.4 M (about 3.7%) glycerol for 1 h at room temperature. They are then transferred to pure Ringer's solution, and after 10-15 min the sarcotubular system of the muscle becomes vacuolated — a phenomenon first described by Krolenko (1967) — with rupture of the connection between the sarcolemma and tubules of the T-system, and hence with the sarcoplasmic reticulum and, through it, with the myofibrils. Only 2% of the elements of the T-system remain connected with the extracellular space. The effect is reversible: the vacuolation disappears if the muscle is transferred back to Ringer's solution with 0.4 M glycerol and the connection is restored both morphologically and functionally.

The neuromuscular effect can be abolished by the addition of 1×10^{-5} g/ml curare to the glycerinated solution and to the pure Ringer's solution, and the action potentials can be blocked by 1×10^{-7} g/ml tetrodotoxin; in this way the contraction which sometimes arises when the muscle is transferred to glycerol-free Ringer's solution is prevented. These precautions are optional.

PREPARATION OF ISOLATED PROTOFIBRILS

(After Huxley, 1963.) The principle of this method is fine homogenization of muscle fibers in a medium with ATP but without Ca^{++}, in which the bond is broken between the myosin and actin protofibrils, so that they can be isolated. Huxley (1963) used EDTA to remove Ca^{++}. A compound with higher affinity for Ca^{++} than EDTA is now available, namely EGTA, and this is recommended for use in this method instead of EDTA.

The psoas muscle of a recently killed rabbit or the pectoral and femoral muscles of a chicken, measuring about 4×2 mm in cross section and 5-10 cm in length are tied at their ends with woollen thread to a transparent plastic slab and immersed in ES-1:

50 volumes glycerol + 40 volumes water + 66 mM phosphate buffer, pH 7.0, at 4°C. Water for all solutions and glycerol are deionized to remove ions of heavy metals. After 24 h the ES-1 is changed at 4°C, and after a further 24 h it is again changed. The product is stored for not less than 3 weeks at $-20°C$. Part of the bundle weighing 0.3-0.4 g is transferred into ES-2: 15 volumes glycerol + 85 volumes of a salt solution consisting of 0.1 M KCl, 1 mM $MgCl_2$, and 66 mM phosphate buffer, pH 7.0. Glycerol is washed away with salt solution of the same composition. The bundles are cut up with scissors into segments about 2 mm in length and homogenized 3 times, for 20 sec each time, with intervals of 15 sec to allow cooling of the homogenizer, which is kept in ice. The fiber is teased into myofibrils, and the result is checked under the phase-contrast microscope. After centrifugation for 3 min at 650 g the residue is resuspended in a small volume (about 1 ml) of relaxing solution at 0°C. The relaxing solution (salt solution + 1 mM EGTA + 10 mM $MgCl_2$ + 3-5 mM ATP, pH 7.0) is added to a volume of 7 ml. Homogenization is carried out immediately, as at first. After centrifugation the supernatant contains isolated protofibrils. The preparation remains stable at 0°C for 24 h, after which aggregation takes place, but this can be prevented by the addition of ATP at the rate of 1 mg/ml suspension with vigorous shaking.

PREPARATION OF ISOLATED SARCOLEMMA SUSPENSION*

PREPARATION OF RELAXING GRANULES FROM SKELETAL MUSCLES

(After Weber et al., 1963.) Minced rabbit muscles are homogenized in a Waring blender for 40 sec at 0°C in 3 volumes 0.12 M KCl in 5 mM histidine buffer (pH 6.5 at 25°C, giving pH 7.0 at 0°C). By centrifugation twice (at 3000 and 8000 g) the myofibrils and mitochondria are removed. Centrifugation is then carried out at 20,000 and 36,000 g for 1 h. The lower, blood-red layer of residue (mitochondria) is then discarded and the upper, white, loosely

---

*See Notes Added in Proof, page 174.

packed layer is collected and resuspended, and then washed twice by dilution 3-5 times with 0.12 M KCl in 1 mM histine buffer, pH 6.5. After centrifugation at 89,000 g the final residue is suspended in the same medium in the proportion giving a suspension which contains 1.2-2 mg N/ml (1 mg N corresponds to 9 mg dry weight of relaxing granules washed with water). The maximal storage life of the granules is 3 days. Before use for experimental purposes the suspension is centrifuged for 7 min at 8000 g to remove the residue storage.

CELL (INTERPHASE) MODELS BECOMING ROUNDED
IN THE PRESENCE OF ATP

(After Hoffmann-Berling, 1953, 1954a.) Objects: 1) tissue cultures of fibroblasts from the amnion and sclera, of the liver and spleen, of the subcutaneous connective tissue, of skeletal and heart muscles and osteoblasts of chickens; 2) fibroblasts from human lymphogranulomatous nodules; 3) epithelial cells from the chicken amnion, retina, and epidermis (the cells must be unconnected); 4) ascites cells of an Ehrlich mouse carcinoma maintained in cell culture.

Tissue cultures on coverslips are convenient. Tissues taken from 8- to 9-day chick embryos are used, but tissues cultivated in vitro for long periods and subcultured many times can be used equally successfully. Before preparation of the models the cultures are kept for 24-48 h in a medium consisting of 25% chick blood plasma, 25% embryonic extract, and 50% Tyrode solution. Embryonic extract is prepared from the tissues of a 9-day chick embryo, homogenized in an equal volume of Tyrode solution and centrifuged. The extract is allowed to stand for 20 min at 37°C, after which the tissue is removed by further centrifugation.

Material for anaphase and telophase models (see below) is prepared in the same way.

Extraction: the objects cooled to 0°C are immersed (in the case of tissue cultures, actually on the coverslips on which they grew), in ES, which is also cooled to 0°C: 0.12 N KCl (0.01 M phosphate buffer, pH 7), 0.004 M cysteine, 45% glycerol. It is better to replace the cysteine by 0.005 M EDTA, when the KCl concentration must be reduced to lower the ionic strength to about 0.15. Some deviation from $\mu = 0.15$ is acceptable. Extraction at

0°C lasts for a few hours to many days. Extraction for several days gives a better contraction in response to ATP.

Washing: in the cold with solution of $\mu = 0.12\text{-}0.16$, buffer not precipitating Ca^{++} and Mg^{++}, pH 7, $2 \times 10^{-2}\,M$ cysteine, and $1.5 \times 10^{-3} - 1 \times 10^{-2}\,M$ $MgCl_2$. Washing is carried out in stages.

The RS is based on the WS, to which $(1\text{-}5) \times 10^{-3}\,M$ ATP is added. Reactivation takes place at room temperature and observations are made under the phase-contrast microscope. The models are conveniently placed in a glass chamber with facilities for the inflow and outflow of the solution, and the chamber itself is placed on the microscope stage.

ANAPHASE MODELS OF FIBROBLASTS

(After Hoffmann-Berling, 1954b.) Objects: tissue cultures on coverslips obtained from the eyelid of an 8-day chick embryo (subcutaneous fibroblasts), which has just appeared at that age. Cultivation and preliminary treatment of the cells are carried out in the same way as before preparation of interphase models (page 146). The cultures are grown for 48 h. To increase the number of mitotically dividing cells the cultures are kept for many hours at room temperature, and then for 1-4 h at 37°C.

Variations in the composition of the solutions used to prepare models are possible and, indeed, sometimes essential. However, it must be remembered that with an increase in the duration and temperature of extraction, and with a decrease in the ionic strength and an increase in the pH of the extracting solution, the model cells become more rigid. This means that the ATP concentration and ionic strength of the RS must be increased and its pH reduced.

In the optimal situation ES has $\mu = 0.15$, and adjustment to this value is made with KCl, 0.01 M phosphate buffer, pH 7.4, $2 \times 10^{-3}\,M$ EDTA, and 50% glycerol. Extraction continues for 14-20 h at $-15°C$ for 3 h at 0°C. Cultures on coverslips are kept for a few minutes at room temperature and then placed in ES cooled to $-15°C$. Rinsing to remove this solution is carried out at 0°C. The coverslips with the model cells are passed through a series of solutions with gradually diminishing glycerol concentration. Eventually they must be placed in a solution with $\mu = 0.16$, 0.01 M phosphate buffer, pH 7.4, and $1 \times 10^{-2}\,M$ $MgCl_2$ (a high Mg^{++} concentration prevents the chromosomes from swelling induced by the plasticizers). The preparation kept in this solution in a chamber fitted with inlet and outlet is examined under the phase-contrast microscope, cells with nuclei in the anaphase stage are sought, and by means of an ocular-micrometer the distance between homologous sets of chromosomes is measured. A solution of the same composition, but containing 0.8 M urea, is then introduced into the chamber. During all these procedures the preparation must remain at its original temperature of 0°C. The models remain in this solution for 5 min, after which the same solution, but containing $(2-3) \times 10^{-3}$ M ATP, is introduced into the chamber. The distance between the sets of chromosomes is then measured again. The measurements are repeated after the model has been washed to remove ATP and urea, so that the nonspecific lengthening of the spindle due to swelling can be allowed for. If the preparation is not treated with urea, the ATP concentration in the RS must be increased to 4×10^{-2} M.

MODELS OF CLEAVAGE IN THE SEA URCHIN EGG

(After Kinoshita and Yazaki, 1967.) Objects: eggs of the sea urchins Pseudocentrotus depressus and Clypeaster japonicus. Immediately after fertilization the eggs are treated for 2 min with 1 M urea solution, after which they are washed with calcium-free sea water. This treatment prevents the formation of fertilization and hyaline membranes and in this way facilitates the extraction. When the eggs in the calcium-free sea water commence the anaphase of cleavage, they are extracted.

The composition of the final ES is as follows: 50% glycerol, 8×10^{-2} M KCl, 1×10^{-2} M MgCl$_2$, 1×10^{-2} M Tris-HCl buffer, pH 7.4. However, the eggs must not be placed at once in this solution, otherwise they would quickly lose large amounts of water and would contract sharply. The following procedure is therefore adopted. Into a 500-ml glass cylinder (5 cm in diameter, 25 cm high) is poured 100 ml ES of the above composition. Next, with great care so as to form separate layers, 50-ml portions of ES containing 40, 30, 20, and, finally 10% glycerol, respectively, are poured in succession on the top. This cylinder is allowed to stand overnight in the cold to allow a glycerol concentration gradient along the height of the cylinder to be established.

The eggs which are now in a certain stage of cleavage are carefully placed in the top layer of the extracting solution, i.e., in the layer containing 10% glycerol. They then sink slowly by the action of their own weight, and pass along the gradient of increasing glycerol concentrations and undergo extraction. For the eggs to sink completely under these conditions down to the bottom layer of ES, which contains 50% glycerol, requires about 8 h. Extraction is taken as complete 6-8 h after its beginning, when the models can be reactivated. Kinoshita and Yazaki do not state at what temperature the extraction should take place. If it is necessary for the models to be kept for a long time they are transferred into a fresh portion of solution containing 50% glycerol and stored at $-14°C$.

Some cells do not fall to the bottom but remain suspended around the half-way mark. These cells are imperfect and must be discarded.

To reactivate the "model eggs" they are placed in solution containing 5×10^{-3} M ATP + 50% glycerol, in which they are kept for 20-30 min at room temperature; meanwhile they are examined under the microscope.

The following point must be remembered. As they pass through the upper layers of the ES, i.e., the layers with a low glycerol concentration, the eggs continue to develop, and development ceases only when the eggs are about half-way down the ES. Apparently, therefore, cleavage was slightly further advanced in the models than in the original material.

These workers stress that it is more difficult to obtain cleavage in models of sea urchin eggs than anaphase division in model fibroblasts, and that in general, work with model eggs is not successful.

TELOPHASE MODELS OF FIBROBLASTS

(After Hoffmann-Berling, 1954c.) Objects: 46 h (or longer) tissue cultures of freshly explanted eyelids of 8- to 10-day chick embryos (subcutaneous fibroblasts) on coverslips. It is very important to use cells of the same batch of explant in any one experiment so that cultivation of all specimens will have reached the same stage. The cultures are prepared just as in the case of interphase and anaphase models (see pages 146 and 147).

ES: 50% glycerol, 1×10^{-3} M EDTA, 1×10^{-2} M phosphate buffer (in February and March, KH_2PO_4:Na_2HPO_4 = 45:55, in other months 35:65), KCl to give $\mu = 0.15$. Extraction at 0°C for 60-90 min. Immersion in ES and transfer to BS must be carried out gradually.

WS: 1×10^{-2} M phosphate buffer (KH_2PO_4:Na_2HPO_4 = 15.85), 1×10^{-2} M $MgCl_2$, KCl to $\mu = 0.15$. Rinsing in the cold.

RS: The same composition as BS but with the addition of ATP, the optimal concentration of which is 2.5×10^{-3} M.

The material is examined under the phase-contrast microscope. It is convenient to place the specimen in a glass chamber fitted with inlet and outlet tubes.

Before addition of the RS, after its addition, and after rinsing the specimen to remove ATP the length of the telophase models and their width in the zone of the equatorial constriction are measured by means of an ocular-micrometer.

MODELS OF MITOCHONDRIA

By the method of Nakazawa (1964). Object: rat liver mitochondria suspended in 0.25 M sucrose. The mitochondria are washed twice with this solution and the material checked by examination in the electron microscope. Final medium: 0.125 N KCl and 0.02 M Tris-HCl buffer, pH 7.4. One milliliter of suspension containing mitochondria from 2 g fresh liver is added to 9 ml of the following solution: 50% glycerol, 0.125 N KCl, and 0.02 M Tris-HCl buffer, pH 7.4. The mixture is kept at -15°C for 3-10 days. Before the experiment it is centrifuged at 26,000 g for 30 min, then washed twice with 0.125 N KCl with the same buffer to remove glycerol. The degree of contraction of the mitochondria by the action of ATP is measured and assessed from the change in optical density of the suspension. To determine the original optical density 0.5 ml of a suspension of mitochondria in 2.5 ml of a mixture containing 0.125 N KCl + 0.02 M Tris-HCl buffer is added to the spectrophotometer cell, mixed, and measured photometrically. The determination is carried out at room temperature, at a wavelength of 520 nm. To the cell containing 3 ml of this suspension is then added 0.1 ml of concentrated ATP and $MgCl_2$ solution to give a final concentration of ATP 5×10^{-3} M and $MgCl_2$ $2 \times 10^{-3} M$ (the calculations must make allowance for the fact that through dilution the concentration of the mitochondria is reduced by 3%).

Several determinations of optical density are made; contraction of the mitochondria reaches a maximum after 3 min.

By the method of Kazakova (1964). Objects: liver mitochondria of mice and rats, mitochondria of rat hepatoma and of Ehrlich's ascites carcinoma. A suspension of mitochondria in 0.12 N KCl with K-phosphate buffer, pH 7.0, was diluted 2-3 times with 50% glycerol and allowed to stand overnight in the cold. The change in optical density at 520 nm under the influence of 2.5×10^{-3} M ATP and 1×10^{-3} M MgCl$_2$ was measured. Swelling of the mitochondria, an effect opposite to that described by Nakazawa and to that observed in fresh, "living" mitochondria, was observed. Admittedly, addition of 2.5×10^{-3} M CaCl$_2$ leads to contraction of these swollen models.

MODELS OF CHLOROPLASTS

(After Parker and Young, 1965.) Object: spinach chloroplasts in 0.3 M NaCl + 2 mM EDTA, pH 7.5. One volume of the suspension is mixed with 9 volumes ES: 50% glycerol, 0.125 N KCl, 0.02 M Tris-HCl buffer, pH 7.5. Extraction at $-15°C$ for 3-10 days. Before the experiment the material is washed twice with centrifugation to remove glycerol and resuspended in 0.35 M NaCl, 0.002 M EDTA, 0.04 M Tris-HCl buffer, pH 7.5. Contraction recorded as an increase in optical density at 540 nm is induced by the addition of 5×10^{-3} M ATP and 5×10^{-3} M MgCl$_2$ to the solution containing: 0.125 N KCl, 0.02 M Tris-buffer, pH 8.0, at 23°C. It reaches its maximum after 12 min.

MODELS OF LOCUST AND FROG SPERMATOZOA

By the Method of Hoffmann-Berling (1955). Object: giant spermatozoa of the locust Tachycines. Diluting solution for the spermatozoa: 10% chick plasma (blood taken from the carotid artery into a waxed tube without anticoagulants) and 80% Tyrode solution, and before the films are made 10% diluted (1:3) chick embryonic extract is added to coagulate the medium (the method of preparation of the chick embryonic extract is described in the section on interphase models, see page 146). The testes are minced in this solution and films are made from the suspension on coverslips which are immersed in ES.

ES: KCl to give $\mu = 0.15$, 1×10^{-2} M phosphate buffer, pH 7.0, 2×10^{-3} M EDTA, and 50% glycerol. Extraction for 24 h at between 0 and $-5°C$, after which the models can be stored in ES at $-18°C$.

WS: KCl to give $\mu = 0.15$, $1 \times 10^{-2} M$ phosphate buffer, pH 7.0, $5 \times 10^{-2} - 1 \times 10^{-1} M$ $MgCl_2$, 2×10^{-3} M cysteine.

RS: the same composition as WS with the addition of 1×10^{-3} M ATP. Reactivation is best carried out at 0°C for 30 min, after which the result is observed at room temperature.

By the method of Ivanov et al. (1959). Object: frog spermatozoa. The testes of spring frogs are minced, treated with 2-3 volumes 50% glycerol, and allowed to stand for 24 h in a refrigerator. One drop of suspension is mixed with 2-3 drops of solution: 0.12 M KCl, 1×10^{-2} M phosphate buffer, pH 7.0, 5×10^{-3} M $MgCl_2$; after a few minutes a drop of ATP is added to the suspension on a slide. The tails of some spermatozoa begin to wave.

MODELS OF MAMMALIAN SPERMATOZOA

(After Bishop and Hoffmann-Berling, 1959.) Object: rabbit, human (by ejaculation), guinea pig, rat, mouse, and bovine (epididymal) spermatozoa. Bovine spermatozoa obtained by flushing out the testis with Tyrode solution injected from a syringe through a needle introduced into the vas deferens are the most convenient object.

Extraction: a suspension of spermatozoa in Tyrode solution (from 1:4 to 1:9) is cooled to 2°C for 1 h after which cold ES is added. The ES contains: 0.12 M phosphate buffer, pH 7, 0.005 M EDTA, 0.006 M $MgCl_2$, 0.001 M thioglycollate, 0.01% digitonin, and 3.5% polyvinylpyrrolidone (this can be replaced by 1% agar). Extraction proceeds for a short time (several minutes or tens of minutes); it must completely immobilize the spermatozoa without disintegrating them.

For reactivation a small quantity of suspension is added to a solution: 0.16 N KCl, buffer not binding Ca^{++} and Mg^{++}, pH 7, 0.004 M $MgCl_2$, 0.001 M $CaCl_2$, 3.5% polyvinylpyrrolidone, and 1×10^{-3} M ATP. Reactivation is carried out at room temperature and observations made under the phase-contrast microscope.

For storage the models are transferred to a solution containing 50% glycerol in which they can be kept at $-20°C$ for several months.

MODELS OF SPERMATOZOA OF CERTAIN MARINE INVERTEBRATES

<u>Cuttlefish sperm</u> (Bishop, 1958b). The sperm of <u>Loligo pealii</u> are extracted for 24 h at 0°C in 50% glycerol, 0.25 N KCl in phosphate buffer (the pH is not specified). They are then transferred to a refrigerator at $-30°C$ in which they can be kept for many days. After rinsing the models are reactivated by solution containing 1×10^{-3} M ATP and 1×10^{-3} M MgCl$_2$ at a temperature not exceeding 12°C.

<u>Sea urchin and starfish sperm</u> (Kinoshita, 1958, 1959). Sperm of <u>Hemicentrotus purcherrinus</u> and <u>Asteria amurensis</u> are extracted for 2-10 days at 0°C in 100 volumes 50% glycerol containing 0.15 N KCl and 0.01 M phosphate buffer, pH 7.5. The extracted tails are concentrated by centrifugation, washed twice with 0.15 N KCl in 0.01 M phosphate buffer, pH 7.5, and reactivated. Besides 0.15 N KCl the reactivating solution must contain 1×10^{-3} M ATP, 1×10^{-2} M MgCl$_2$, and 1×10^{-2} M histidine. The pH of the solution must be 7.2-7.8. Instead of histidine other chelating agents (cysteine, versene, thioglycollate) can be used.

MODELS OF CILIATED EPITHELIUM

<u>By the method of Aleksandrov and Arronet</u> (1956). The palatal mucous membrane and "uterine" part of the oviduct of frogs are the most convenient objects. In the first case the soft tissues of the palate are excised, the film is cut into 6-10 pieces with a razor, and these are immersed in cold ES of the following composition: 0.12 N KCl, Tris-maleate (or other) buffer, pH 7, and 45% glycerol. Extraction proceeds in the cold for 24 h or more. Rinsing with a solution: 0.12 N KCl, buffer not precipitating Mg^{++}, pH 7, and 5×10^{-3} M MgCl$_2$. After rinsing to remove glycerol, the mucous, consisting of macerated ciliar and mucous cells, is removed from the piece of tissue on a slide and 1-2 drops of the following solution are added: 0.12 N KCl, buffer pH 7, 5 ×

10^{-3} M MgCl$_2$, and $5 \times 10^{-3} - 1 \times 10^{-2}$ M ATP. After mixing the suspension of cells is covered by a coverslip. Examination should be carried out under the phase-contrast microscope, but an ordinary water-immersion objective or even a dry objective giving magnification of 40× or 20× can be used.

Pieces of the lower part of the oviduct are treated in the same way except that scrapings are not made from them because extraction does not lead to maceration of the epithelium. The edge of the piece of tissue, where a layer of ciliated cells can be seen, is examined under the microscope. Naturally the phase-contrast microscope cannot be used in this case because the piece of tissue is too thick.

Models can also be made from ciliated cells of the frog palate by means of an ES containing saponin and not glycerol. The saponin concentration is 0.1-0.2% and the extraction time 30-90 min, but since saponin is a collective term for a whole group of different substances, the proper concentration and extraction time must be chosen individually for each preparation. An inconvenience of saponin extraction is that, as a rule, the conversion of the cells into models is insufficiently synchronized: when most cells have become models, some of them still retain their spontaneous oscillation. Lengthening the extraction time or increasing the saponin concentration converts these "laggard" cells into models but at the same time causes the great majority of models to be completely dissolved.

Glycerol extraction can be used to obtain models of the tracheal ciliated epithelium of mammals and some reptiles. The trachea is cut transversely into thin rings which are extracted, rinsed, laid flat on a slide or coverslip in reactivating solution, and the inner surface of the ring is examined under a microscope, preferably with a water-immersion objective.

Extraction by detergents has not been tried on tracheal epithelial cells. Perhaps it would yield good results.

By the method of Child and Tamm (1963). Object: ciliated cells of the branchial plates of the mussel Modiolus demissus.

The excised branchial plates are extracted at 0°C for about 1.5 h in ES: 40 vols.% ethylene glycol, 10% glycerol, 0.01 M KCl, 0.01 M phosphate buffer, pH 7.5.

The extracted plates are rinsed at 0°C for 2 min in **WS:** 0.1 M KCl, 0.005 M MgCl$_2$, 0.01 M histidine, and 0.01 M phosphate buffer, pH 7.5.

The rinsed plates are reactivated in RS consisting of BS with the addition of 3×10^{-3} M ATP, initially at 0°C and later with gradual heating of the preparation to 23-26°C. Observations are made under the phase-contrast microscope.

MODELS OF THE VORTICELLA STALK

By the method of Levine (1956). Objects: Vorticella campanula, V. nebulifera, and V. convallaria.

The **vorticellas** are washed in distilled water and immersed in cold (0°C) 20% or 50% glycerol solution containing 4×10^{-3} M EDTA (to prevent coiling of the spasmoneme) and with pH 7. They can be kept in this solution at 0°C for not more than 1 month (depending on the glycerol concentration and species of **vorticella**).

Glycerol is rinsed off with a solution containing 0.5 N KCl and 4×10^{-3} M EDTA. Contraction of the stalk proteins is induced by the presence of 3×10^{-3} M CaCl$_2$ in this solution, and uncoiling by transferring them back to solution containing 4×10^{-3} M EDTA; according to Levine ATP has no action on the stalks.

By the method of Hoffmann-Berling (1958). Object: V. gracilis. Pieces of cotton wool to which the vorticellas attach themselves are first placed in the culture medium (a suspension of yeast). Together with the cotton wool they are immersed in ice-cold lytic solution of the following composition: 0.12 N KCl, 0.02 M Tris-maleate buffer, 4×10^{-3} M EDTA, 0.1% saponin, pH 6.8. They are kept in this solution for 20-30 min, and then either reactivated experimentally or placed in preserving solution (40-50% glycerol, 0.12 N KCl, 0.01 M phosphate buffer, 4×10^{-3} M EDTA, pH 6.8).

The models are reactivated in the following solution: 0.12 N KCl, 0.02 M Tris-maleate buffer, pH 6.8, 1×10^{-5} g-ion/liter Ca^{++}. The models are relaxed by 1×10^{-3} M EDTA. The stalk can be made to contract rhythmically by placing the models in a solution: 0.12 N KCl, phosphate buffer, pH 6.6-6.8, $(1-4) \times 10^{-3}$ M ATP, $(2-4) \times 10^{-3}$ M MgCl$_2$, and if a freshly prepared model is used, 1×10^{-4} M CaCl$_2$. The higher the ATP and Mg^{++} concentrations, the higher the Ca^{++} concentration required. Old models require no Ca^{++} whatsoever.*

By the method of Seravin (1963). Two methods of extracting the Vorticella stalk are available (the colonial vorticella Carchesium polipinum is used). 1) The vorticellas are immersed in ES: 0.12 N KCl, 4 × 10^{-3} M EDTA, 0.01 M phosphate buffer, pH 7, after which they are frozen in it with liquid air. The same ES at room temperature is added, after which the colony is quickly transferred into ES cooled to 0°C, in which it is kept for between 5 min and 4 h. 2) Microaquaria containing vorticellas in ES of the same composition are placed for 5 min in a water thermostat at t = 50°C, then quickly transferred to the same ES cooled to 0°C, in which they remain at the same temperature for about 6 h.

Both methods — freezing and rapid heating — cause disruption of the membranes and facilitate subsequent extraction. After extraction, the models are transferred to a solution without EDTA. Contraction of the branches is induced by 5 × 10^{-5} (or higher) M $CaCl_2$, and relaxation and lengthening by ATP + $MgCl_2$. A combination of 1 × 10^{-3} M ATP + 1 × 10^{-3} M $CaCl_2$ induces rhythmic alternation of contraction and lengthening of the branches.

MODEL CILIA FROM INFUSORIANS

Personal observations. The infusorians (Spirostomum ambiguum) are placed in cold ES: 0.3-0.12 N KCl, 0.001-0.01% digitonin, 0.004 M EDTA, phosphate buffer, pH 7. The extraction time (and also the digitonin concentration) must be chosen individually and the optimal time is between 15 and 120 min. Models can be kept after extraction in a solution containing 40% glycerol for several hours. The models are rinsed to remove ES by buffered (pH 7) 0.12 N KCl. The reactivating solution contains 0.03-0.12 N KCl, 2 × 10^{-3} M ATP, and 3 × 10^{-3} M $MgCl_2$; it is buffered at pH 7.0.

Addition of Ca^{++} to the medium stimulates oscillation of the cilia, but also leads to total lysis and disintegration of the infusorians. Generally speaking the bodies of model infusorians of this genus are easily disintegrated by any mechanical agency, and these models are therefore best examined under the microscope in a

---
*See Notes Added in Proof, page 175.

hanging drop, so that the phase-contrast microscope can be used provided that a thin slide with a well is available.

Spirostomum models have also been obtained with the aid of 0.1% saponin as lytic agent, but the percentage of cells converted into models is smaller when this technique is used, and more of them are destroyed.

To make models from Paramecium caudatum the infusorians are placed in cold ES: 0.0005% digitonin, 0.03 N KCl (or without this component), 0.01 M phosphate buffer, pH 7, and 0.004 M EDTA. After extraction for 2-3 h the ES is rinsed out with buffer or 0.03 N KCl in buffer, pH 7. RS: 2×10^{-3} M ATP, 1×10^{-4} M CaCl$_2$, 3×10^{-3} M MgCl$_2$, and 0.03 N KCl (or without it), Tris-maleate buffer, pH 7. For the preparation of Paramecium models Hoffmann-Berling recommends the use of milder lytic agents — cationic detergents (for example, Tween preparations). He also recommends that an excess of Mg^{++} (6×10^{-3} M) be added to the ES containing EDTA in the ES so that Ca^{++} is removed from the solution but Mg^{++} remains in it. (It would probably be better if EGTA were used for this purpose.)

Paramecium models are highly unpredictable objects. For each culture the digitonin concentration in the ES and the extraction time must be chosen individually. Models have so far been obtained only from rice cultures but not from Paramecium cultures grown on a salt medium.

By the method of Seravin (1961, 1967). To prepare models from Spirostomum ambiguum the use of a solution with a higher saponin concentration (0.5%) and pH 6.5-6.6 is recommended, or that a saturated solution of digitonin be used. Seravin (1961) describes a method of preparing models from Euplotes patella: the infusorians are rinsed in bidistilled water and placed in ES at 0°C; 0.12 M KCl, 0.01 M phosphate buffer, pH 6.9-7.0, and 0.15% saponin or saturated digitonin solution. Extraction proceeds in the cold for 20-45 min, after which the product is rinsed with detergent-free solution and reactivated by solution containing 1×10^{-3} M ATP and 1×10^{-3} M MgCl$_2$.

Preparation of Paramecium models with reversed ciliary beating. Naitoh (1969) has developed a method of preparing models of Paramecium caudatum whose cilia can be reoriented toward the anterior end of the cell (reversal of beating) under the influence of ATP, Ca^{++}, and Zn^{++}. The infusorians on an infusion of hay are rinsed with 2 mM $CaCl_2$ in 10 mM Tris-HCl buffer, pH 7.4. The suspension is cooled to 0°C and centrifuged to obtain irregularly spherical bodies from the cells. They are resuspended at 0°C in extracting solution: 50% glycerol, 50 mM KCl, 10 mM EDTA, 10 mM Tris-HCl buffer, pH 7.4. The models are kept at $-15°C$ for 10-15 days in this solution. They are then washed 3 times with 50 mM KCl made up in the same buffer, at 0°C. The models are kept in this solution for not less than 15 min before the experiment. A suspension with the minimal quantity (about 1×10^{-4} ml) of solution is placed in reactivating solution (volume about 1 ml) of the following composition: 10 mM ATP, 10 mM $CaCl_2$, 0.1 mM $ZnCl_2$. Naitoh recorded the change in orientation of the cilia by photographing models before and after reactivation.

Models of isolated infusorian cilia (After Gibbons, 1965). Object: Tetrahymena pyriformis cultivated on medium: 1% peptone, 0.1% yeast extract, phosphate buffer, pH 6.5.

The cells are concentrated by centrifugation, washed in solution containing 0.18% NaCl and 0.2 M sucrose, then resuspended in fresh solution up to a volume of 20 ml; cells account for half this volume. The concentrated suspension is cooled to 0°C and mixed with 100 ml ES: 70% glycerol, 50 mM KCl, 2.5 mM $MgSo_4$, and 20 mM Tris-thioglycollate buffer, pH 8. The suspension is then cooled to $-20°C$ and kept at this temperature. The suspension is then treated vigorously in a Vortex mixer for 1 min. As a result most of the cilia are separated from the cell bodies, but the cell bodies themselves remain intact. The bodies are precipitated by centrifugation for 10 min at 12,000 g. The supernatant, a suspension of pure cilia, can be kept at $-20°C$. However, the suspension is best used in reactivation experiments on the day it is made, for during storage even for a few days the proportion of cilia capable of reactivation falls off considerably.

The suspension of cilia is reactivated by mixing it with RS: 0.2 M ATP, 50 mM KCl, 2.5 mM $MgSo_4$, and 20 mM imidazole-HCl buffer, pH 6.8, at 20°C. The optimal effect is obtained by using suspension and RP in the ratio of 1:3 by volume. The mixture of suspension and RP must be examined quickly under the phase-contrast microscope at room temperature.

Material for preparing models of isolated cilia is best obtained not by grinding the infusorians but by treating them with ethanol-calcium solution (Watson and Hopkins, 1962). The technique of cultivation is described by Watson and Hopkins (1962): 8 liters of culture is sedimented by centrifugation for 10 min at 1000 rpm, the supernatant is quickly decanted, and the residue is shaken in 800 ml water and centrifuged for 10 min at 100 g and 4°C, and again washed by resuspension in 800 0.025 M sodium acetate, pH 7.0. Next, 750 ml of a solution of the following composition is added: 12% ethanol, 0.025 M sodium acetate, 0.1% EDTA (the trisodium salt, pH adjusted to 7.0 with acetic acid), and then 25 ml of 1 M $CaCl_2$ is quickly added. At this stage the temperature of the suspension is about 8°C and its pH 5.8. The suspension is allowed to stand for 10 min in a cold room and shaken from time to time. The cilia separate from the cells. The cell bodies are removed by centrifugation for 10 min at 1000 g and 4°C. Part of the supernatant (600 ml) is drawn off carefully. The rest of it is not removed so as not to disturb the residue and to contaminate the isolated cilia with cell bodies. For the same purpose the centrifuge is started up and slowed down very gently. If despite everything the residue is stirred up, further centrifugation will not remove traces of the cell bodies. If that happens, the suspension must be filtered (without suction) through a porous No. 3 glass filter. The same technique can be used also to isolate cilia from a small volume of culture of Tetrahymena pyriformis. A volume of culture containing about 1×10^6 cells is taken. The cells are sedimented by centrifugation and washed, and the residue is treated with 0.6 ml 0.025 M sodium acetate and 3.0 ml 0.1% EDTA, followed by 0.1 ml 1 M $CaCl_2$. The isolated cilia can be extracted in order to prepare the models. Winicur (1967) extracts cilia isolated by ethanol-calcium treatment either with glycerol-containing solutions or with solutions of 0.05% digitonin with 60% sucrose, or with ethylene glycol solution.

MODELS OF FLAGELLA OF UNICELLULAR ALGAE

(After Brokaw, 1960, 1961, 1963.) Object: Chlamydomonas moewusii. A culture grown on mineral medium is sedimented by centrifugation, washed with 0.02 M Tris-thioglycollate buffer, pH 7.8, and placed in cold ES (3-4 volumes; 0°C): 0.01 M Tris-thioglycollate buffer, pH 7.8, 0.01 M $MgCl_2$, and 70% glycerol. The sample is rapidly cooled to $-20°C$. After 1 h the suspension is shaken vigorously and centrifuged for 15 min, at 10,000 g. The cell bodies settle while the flagella remain in suspension. The sample of supernatant is diluted 1:10 with the following solution: 0.05 N KCl, 0.004 M MgCl$_2$, 2×10^{-4} M ATP, and 0.02 M Tris-thioglycollate buffer, pH 7.8. Observations are made in a dark field at 10-15°C with magnification 560×.

The same technique can be used for the unicellular alga Polytoma uvella with the exception of the two following recommendations: shortening the extraction time to 30 min (Brokaw, 1961) and adding 2.5% polyvinylpyrrolidone to the reactivating solution (Brokaw, 1963).

MODELS OF FLAGELLA FROM PROTOZOANS

By the method of Hoffmann-Berling (1955). Objects: Trypanosoma brucei and T. gambiense. Blood of mice infected with typanosomes is mixed in the ratio of between 1:20 and 1:40 with the following solution: 10% chick plasma (blood taken from the carotid artery into a waxed tube without anticoagulants), 80% horse serum, and with the addition, just before making the films, of 10% dilute (1:3) chick embryonic extract (to coagulate the medium); for the method of preparing the extract, see page 146). Films are made. After the fibrin has clotted (in a humid chamber) the slides are dipped in ES. The composition of the solutions and all the rest of the technique are the same as for models of locust sperm (see page 151).

By the method of Seravin (1967). Objects: The flagellates Peranema trichophorum and Chilomonas paramecium. ES: 0.12 M KCl, 0.01 M phosphate buffer, pH 7, 0.004 M EDTA, the solution is saturated with digitonin. Extraction time 30-45 min, temperature around 0°C. RS: 0.12 M KCl, 0.01 M phosphate buffer, pH 7, 1×10^{-3} M ATP, 1×10^{-3} M MgCl$_2$.

MODELS OF AMEBAS

By the method of Simard-Duquesne and Couillard (1962a). Amoeba proteus cells are placed for 10 min in 5×10^{-4} M EDTA at room temperature. The suspension is then gradually taken through a series of extracting solutions of increasing glycerol concentration at 0-4°C. The composition of the

ES is: KCl to give $\mu = 0.05$ or 0.1, $0.01\ M$ Tris-maleate buffer, pH 7, $5 \times 10^{-3}\ M$ EDTA, various concentrations of glycerol (5, 10, 20, 30, or 40%), allowing 30 min in each solution. Extraction at 0-4°C in the last ES, containing 40% glycerol, continues for 24 h and later at $-10°C$ for 3-6 days. The model amebas together with pieces of glass wool are placed on a slide. By means of 3 coverslips (the middle one rests on the two outer ones, which are polished) a small chamber is made into which the glass wool with the amebas is placed. Washing solution is run through the chamber at room temperature: this consists first of a series of extracting solutions with diminishing glycerol concentrations, followed by WS: $0.01\ M$ Tris-maleate buffer, pH 7.0, $5 \times 10^{-2}\ M$ EDTA, KCl to $\mu = 0.05$. Finally, the RS differs from the WS in not containing EDTA, but instead it contains $0.02\ M$ cysteine, $3 \times 10^{-3}\ M$ MgCl$_2$, and $1 \times 10^{-3} M$ ATP.

By the method of Seravin (1967). Cells of A. proteus are introduced into ES containing 25% glycerol and cooled to $-5°C$, after which the tube is immediately and rapidly cooled to $-25°C$. The ES with the amebas freezes. After 10 min the tube is warmed in a stream of water at room temperature. The ES quickly thaws and the tube is allowed to stand at between -1 and $+5°C$ for 2-24 h or more. Freezing ruptures the membranes, after which extraction proceeds. The models are then washed to remove glycerol and reactivated by a solution containing $1 \times 10^{-3}\ M$ ATP and $1 \times 10^{-3}\ M$ MgCl$_2$.

PROTOPLASMIC STREAMING MODELS

IN THE MARINE ALGA Acetabularia calyculus

(After Takata, 1961.) Stalks 2 cm long without caps are used. They are tied with a silk ligature at the base and then separated from the skeleton and placed in $1 \times 10^{-2}\ M$ phosphate buffer, pH 7, containing $0.1\ N$ KCl and $1 \times 10^{-3}\ M$ CaCl$_2$ or $1 \times 10^{-3}\ M$ MgCl$_2$ for 8-10 min at room temperature, then in ES containing 20% glycerol and phosphate buffer, pH 7 (the ES probably should contain $0.1\ N$ KCl also) for 18-20 h (the temperature is not specified). After rinsing, reactivation is carried out with $1 \times 10^{-3}\ M$ ATP and $1 \times 10^{-3}\ M$ CaCl$_2$ or MgCl$_2$. Motility continues for 1-2 min.

MODELS OF THE MYXOMYCETE PLASMODIUM

(After Kamiya and Kuroda, 1965.) Object: the plasmodium of the myxomycete Physarum polycephalum. A small piece of protoplasm of the plasmodium weighing a few milligrams is placed on the surface of a thin film of 4% agar. The plasmodium spreads over the film on which it forms a thin layer. It is then covered with a strip of well-washed cellophane, and this in turn is covered by another agar film 0.4-0.5 mm thick. For some time the plasmodium, inside this "sandwich," continues to slide between the layers of agar and cellophane to form numerous bands, in which intensive streaming of protoplasm can be observed under the low power of the microscope. The upper and lower layers of agar are then sealed at the edges with liquid agar, and as a result a single closed agar block surrounding the thin layer of plasmodium is obtained. This block is immersed in cold ES of the following composition: 40% glycerol, 1×10^{-2} M EDTA, 1×10^{-2} M KCl, 1×10^{-2} M Tris buffer, pH 7.1, and kept at a temperature of between -5 and $-7°C$ for 24 h. The blocks are then transferred to a fresh portion of the same solution and kept for 1-5 weeks at either -15 or $-5°C$.

After the upper layer of agar and the cellophane film have been removed the preparation is washed with several portions of 6×10^{-2} M KCl at 23-25°C. Reactivation is then carried out with one of the following solutions (the results are the same whichever solution is used): 5×10^{-3} M ATP, 5×10^{-3} M MgCl$_2$, and 3×10^{-2} M KCl. 1×10^{-3} M ATP, 3×10^{-3} M MgCl$_2$, 2×10^{-2} M KCl, and 1×10^{-2} M Tris buffer, pH 7.2. The specimen together with the agar backing is placed in a vessel (such as a Petri dish) containing one of the above solutions and examined under the microscope with a 10× objective.

THE USE OF MODELS FOR PRACTICAL WORK WITH STUDENTS

ATP and Muscle Contraction

The frog sartorius or ileo-fibularis muscle is tied at its ends with threads so that it cannot contract (i.e., develop a contracture) on treatment with extracting solution. For this purpose a glass or plastic (but not metal) bracket can be used. The muscle, thus kept under isometric conditions, is placed for about 24 h in ES at between 0 and +4°C (the composition of the solutions is given on page 136). After 24 h, thin bundles, each consisting of 2-5 fibers, are separated from the muscle without removing it from the bracket. The dissection is carried out under a binocular loupe with magnification 8×1 or 8×2, using thin dissecting needles. The needles can be prepared if necessary from entomological pins by fixing them, for example, in a glass tube with Mendeleev's paste; it is useful to sharpen the pins and to flatten them like a scalpel. The bundles are rinsed in BS and placed on a slide, a drop of RS is added, and contraction of the bundle is observed. The slide can be placed on squared paper so that the initial length of the bundle and its length after contraction can be noted. Instead of ATP, muscle juice (see page 136) can be used.

ATP and Ciliary Movement

Observation of the oscillations of cilia on normal ciliated epithelial cells and their immobilization by metabolic inhibitors. The palatal mucous membrane of a frog is excised and cut with a razor into pieces measuring about 4×4 mm. The pieces are placed in a drop of Ringer's solution on a slide and covered by a coverslip. To prevent pressure on the piece of tissue by the coverslip, two strips of filter paper (2-3 layers) or a fragment of coverslip, or pieces of wax are placed between the slide and coverslip to act as supports for the coverslip; in this way a shallow chamber is created, having the slide as its base and the coverslip as its lid. Under a microscope (objective $40\times$, preferably water-immersion, or $20\times$) oscillation of the cilia on the epithelial cells is observed; this is best seen at the edges of the piece of tissue, especially if the ciliated surface of the mucous membrane is underneath the piece of tissue. The pieces are then placed in Ringer's solution containing either dinitrophenol (1×10^{-3} M), sodium azide (5×10^{-3} M), or sodium fluoride (5×10^{-2} M). After 30-40 min the piece of epithelium, which has remained during this time in the solution containing the inhibitor, is again examined under the microscope in a drop of the same solution. Absence of ciliary movement can clearly be observed. The pieces are rinsed with Ringer's solution without inhibitor and allowed to stand in a fresh portion of pure solution. After 15-20 min microscopic examination shows that the cilia are once again in movement. These experiments are highly reproducible. All that must be remembered is that the solution of dinitrophenol, if this is used as the inhibitor, must be freshly prepared.

The object of another exercise is to show that ATP is the source of energy for mechanical work of the cilia and that metabolic inhibitors of the dinitrophenol, sodium azide, and sodium fluoride type do not act directly on the contractile system of the cilia, and that, consequently, they inhibit oscillation of the cilia by acting on a different system.

The first stage is to examine previously prepared models of the ciliated epithelium of the frog palate in the bathing solution under the microscope (for the method of preparing and observing models of ciliated epithelium, see page 153); the cilia are observed to be immobilized. The models are then examined under the microscope in reactivating solution, i.e., in solution containing ATP and Mg^{++}; oscillation of the cilia on the model cells can then be observed, including cilia on fragments of cells or, if the optical system is good, isolated cilia also.

Some of the extracted pieces of palate are placed in bathing solution containing the same inhibitor as in the experiment on living epithelial cells. The incubation time for these pieces in this bathing solution and the concentration of inhibitor in the solution are the same as in the experiment on living cells. At the end of this time the models are reactivated by solution containing not only ATP and Mg^{++}, but also the inhibitor as before. Microscopic examination shows that the cilia of the models oscillate despite contact with the inhibitor, whereas oscillation of the cilia of the living cells under these circumstances is arrested.

* * *

The investigations surveyed in this book originated, on the one hand, in the work of V. A. Engel'gardt and M. N. Lyubimova, the founders of mechanochemistry, and on the other hand, in the work of H. H. Weber, A. G. Szent-Györgyi, and H. Hoffmann-Berling, who added motile muscle and cell models to the scientific armamentarium. The progress made in recent decades in the study of the mechanism of cell motility has largely been due to the efforts of these investigators. The author of this book regards his survey as a testimony of his deep respect for the work and names of these scientists.

The author takes this opportunity of expressing his gratitude to Marina Fedorovna Konstantinova for her invaluable help with the writing of this book and with the work on which it is based, and to Professor Vladimir Yakovlevich Aleksandrov for editing the survey and directing the work. Much valuable advice was given by E. K. Zhukov, S. A. Krolenko, V. I. Vorob'ev, and L. N. Seravin. Without their aid many details would have been overlooked. I am indebted to them all for their help.

Notes Added in Proof

To page 23

During the first days of glycerol extraction the ATPase activity of the fibers is reduced; probably the structure of the sarcolemma at this time is still too compact and it is therefore not readily permeable to ATP and to the reaction medium as a whole. Correspondingly, on the first day ATP does not cause contraction of the fiber. The ATPase activity later increases, and after 6-8 days of extraction it becomes stabilized at the level of 80 μg P/mg protein (Kalamkarova and Veretennikova, 1970). The same workers showed that extracted fibers also retain their cholinesterase activity. It decreases during the first days but increases after the 8th-10th day, and by the 16th-18th day it reaches the level of 30 μg acetylcholine/mg protein, at which it subsequently remains. Kalamkarova and Veretennikova consider that this cholinesterase activity is connected with the myofibrillary proteins themselves and that it plays some sort of direct role in the contractile cycle. Extraction of the sarcoplasmic proteins from the fiber is complete after about 3 weeks (Szent-Györgyi, 1949), after which the model fiber consists mainly of myofibrillary proteins and its composition subsequently remains unchanged.

To page 24

According to Chichibu (Tohoku J. Exp. Med., 73:170, 1961), the muscle fiber completely loses all the electrical properties characteristic of the living muscle during the first few hours of glycerol extraction.

To page 35

Solaro et al. (1971) treated a washed homogenate of dog's heart muscle with Triton X-100. This compound dissolved all the membranous structures remaining in the homogenate, including mitochondria and their fragments and pieces of the sarcolemma membranes and sarcoplasmic reticulum. As a result a preparation of myofibrils free from impurities was obtained. The homogenate, even if well washed with salt solution, unless treated with Triton possesses significant activity of the mitochondrial enzymes cytochrome oxidase and azide-sensitive ATPase, the enzyme ($K^+ + Na^+$)-dependent ATPase bound with the sarcolemma, and quinidine-sensitive ATPase, bound both with the mitochondria and with the sarcolemma. After treatment of the preparation with Triton followed by washing all these types of enzyme activity almost completely disappear. Meanwhile Triton X-100 does not damage the contractile proteins of the myofibrils: the ATPase activity of the pure preparation of myofibrils corresponds exactly to the ATPase activity of cardiac actomyosin.

To page 42

In addition, methods of obtaining isolated sarcolemma have been developed (Kono and Colowick, 1961; Rosenthal et al., 1965; Madeira and Carvalho, 1969). Homogenate of skeletal muscles or myocardium is repeatedly extracted with various solutions dissolving sarcoplasmic and myofibrillary proteins. Nuclei, mitochondria, and fragments of sarcoplasmic reticulum are also removed from the preparation by centrifugation. Ultimately the preparation contains cleanly washed empty transparent tubules 50-100 μ in diameter: fragments of the sarcolemma of the muscle fibers. Some tubules are open at both ends, others are closed at one end – these are from the sarcolemma from the ends of the fibers. The sarcolemma consisted of three layers: the outer layer is formed from collagen fibers, the middle layer (basement membrane) from amorphous material, while the inner layer is the plasma membrane. The outer and middle layers can be destroyed with collagenase; as a result a preparation consisting of pure plasma membranes is obtained (Kono et al., 1964).

The isolated sarcolemma prepared in this way can be used as an object for the study of permeability of the sarcolemma to various substances, its structure, composition, and its enzymic activity.

Madeira and Carvalho (1969) found that the isolated sarcolemma of rabbit skeletal muscle fibers contracts under the influence of 10 mM ATP. The contraction takes place in a radial direction, as a result of which the diameter of the tubules is reduced. This contraction has been shown not to be due to an osmotic effect: hypotonic and hypertonic solutions do not induce it. Nor can it be explained by the activity of remnants of the myofibrillary apparatus left behind after incomplete washing. This conclusion is supported by the following facts. 1) Contraction of the sarcolemma also takes place in a medium containing 1-2 mM ATP + 10 mM EGTA; in this concentration EGTA is known to suppress the contractility of the actomyosin complex, for it binds virtually the whole of the Ca^{++}; it will be noted that EGTA itself does not induce contraction of the sarcolemma but it makes its contractile structures more sensitive to ATP, thereby reducing the threshold concentration of ATP for contraction from 5-10 mM to 1-2 mM. 2) Contraction of the sarcolemma can take place in medium containing ATP + 1 mM KCl; actomyosin is dissolved in such a medium. Contractility is thus a property of the sarcolemma itself and it cannot be attributed to any possible contamination of the preparation.

ADP acts in the same way as ATP on the sarcolemma. AMP does not induce contraction. If, however, 10 mM ATP is added after AMP, the sarcolemma contracts in a different way, namely in the longitudinal direction, i.e., it shortens.

It is not yet clear whether this contraction of the sarcolemma is connected with hydrolysis of ATP. Nor has it been discovered which layer of the sarcolemma is responsible for its contraction. However both these problems are readily open to experimental investigation.

To page 62

Thrombostenin is known to consist of actin-like and myosin-like proteins similar in their properties to the corresponding muscle proteins (Adelstein et al., 1971). An actomyosin-like protein has also been obtained from leukocytes (Senda et al., 1969) and from other objects.

To page 68

Behnke et al. (1971a) found actin-like and myosin-like filaments in platelets. By using the same method, namely, incubation of glycerinized models in medium containing H-meromyosin, they found actin filaments in spermatids, spermatocytes, and mature spermatozoa of insects (Behnke et al., 1971b; Gawadi, 1971). These workers showed that actin filaments about 50 Å in diameter forming arrow-shaped structures with H-meromyosin are arranged in the dividing cells along the long axis of the spindle and also in peripheral parts of the cells. In other words, actin fibers occur in those parts of the cells where motor reactions are taking place.

To page 76

However, in investigations conducted recently in Mazia's laboratory (Mazia et al., 1972; Petzelt, 1972) it was shown that the isolated mitotic apparatus possesses ATPase activity which is activated by Ca^{++}. The specific activity of the ATPase concentrated in the mitotic apparatus is two to three times greater than the specific activity of the ATPase in other parts of the cytoplasm. In addition, other workers have found actin filaments in the spindle of dividing insect spermatids and spermatocytes (Behnke et al., 1971b; Gawadi, 1971).

To page 94

Quantitative data on the effect of Ca^{++} on motor activity of the models and, in particular, on the vorticella stalk obtained before Ca-EGTA buffers began to be used require verification and more exact redetermination using these buffer solutions by means of which a precise Ca^{++} concentration can be produced. As regards the vorticella stalk this has now been done by Amos (1971). He showed on the glycerinized stalk of Vorticella convallaria that a concentration of Ca^{++} of 5×10^{-7} g-ion/liter is the threshold value for motor activity of the stalk. With Ca^{++} concentrations below 1×10^{-7} g-ion/liter the model stalks are relaxed, but on increasing the Ca^{++} concentration above 5×10^{-7} g-ion/liter they contract. By alternating media with 1×10^{-6} and 1×10^{-8} g-ion Ca^{++}/liter, Amos obtained 35 "contraction−relaxation" cycles of the same model stalk.

Amos calculated that the quantity of Ca^{++} required to produce maximal contraction of a single myoneme is 6.9×10^{-17} g-ion. This is equal to the quantity of Ca^{++} contained in 1.1×10^{-3} M $CaCl_2$ solution in a volume equal to the volume of one myoneme.

Various inhibitors (2 mM Salyrgan, 1 mM KCN, 2 mM dinitrophenol, 0.5 mM 2,4-dinitro-1-fluorobenzene) do not prevent contraction of the model stalk in response to the action of Ca^{++}. In experiments with Ca-EGTA buffers, Mg^{++} and ATP had no appreciable action on coiling of the glycerinized stalk or on the value of the threshold Ca^{++} concentration inducing its contraction. However, model stalks contracted by the action of 0.1 mM $CaCl_2$ solution react to the addition of 2 mM ATP + 4 mM $MgCl_2$: some of them are more tightly coiled, others are relaxed and lengthened, after which pulsation begins as described by Hoffmann-Berling.

A detailed investigation of the roles of pH, Ca^{++}, Mg^{++}, and ATP in the initiation of contraction and relaxation of the stalk was carried out on the glycerinized stalk of Vorticella convallaria by Townes and Brown (1965). These investigators found that in a medium of optimal composition (3×10^{-9} g-ion Ca^{++}/liter, 4×10^{-3} g-ion Mg^{++}/liter, 6×10^{-3} M ATP, 0.12 M KCl) the glycerinized stalk is relaxed at pH values of 6.8 or below, but it is maximally contracted with a change in pH from 6.8 to 7.0. If Ca^{++} is completely absent from the medium, the pH threshold below which the stalk is in a relaxed state and above which it is contracted is 7.5. The concentration of Ca^{++} required to lower this threshold pH to 6.8 is no more than 3×10^{-9} g-ion/liter. Under these conditions an increase in the Ca^{++} concentration to 3×10^{-6} g-ion liter leads to contraction of the stalk at pH 6.8. The action of Ca^{++} on the glycerinized stalk is thus twofold: first, it affects its contraction–relaxation cycle induced by changes in pH; second, Ca^{++} can itself induce contraction of the stalk.

ATP and Mg^{++} at pH values above 7.0, together with Ca^{++}, determine the degree of contraction of the stalk; at pH values below 6.8, conversely, they affect the degree of relaxation of the stalk: the higher the ATP and Mg^{++} concentration in the medium the greater the degree of relaxation. Each of these agents (ATP and Mg^{++}) in the different regions of pH values thus produces the opposite effect on the mechanical activity of the glycerinized stalk.

Townes and Brown consider that Ca^{++}, ATP, and Mg^{++} form a complex with the contractile system of the stalk which changes, with a shift of pH, from the relaxed form to the contracted and vice versa. If all components of the complex are present in optimal proportions, its association—dissociation constant is such that the transition to the contracted form of the complex takes place with a change in pH from 6.8 to 7.0. Unlike ATP and Mg^{++}, Ca^{++} is evidently not merely a participant in the contraction process, but is also its initiator, through its influence on the association-dissociation constant of the hypothetical complex.

Favard and Carasso (1965) found that the myonemes of Peritriches, especially the myonemes of their stalks, contain an endoplasmic reticulum. It is regularly arranged in the stalk and alternates with contractile filaments of the myoneme, the tubules of the reticulum being in close opposition to the myofilaments. By using specific cytochemical reactions demonstrating Ca^{++} electron-microscopically, Carasso and Favard (1966) showed that the reticulum of the myoneme is a reservoir of Ca^{++}. A high concentration of Ca^{++} is stored within the tubules of the reticulum and is distributed uniformly in them. Very probably, just as in muscle migration of Ca^{++} from the tubules of the reticulum to the contractile elements provokes their contraction with the resulting contraction of the stalk. The return of Ca^{++} to the tubules can be presumed to lead to relaxation of the stalk.

To page 109

This phenomenon — reorientation of the cilia of Paramecium models in the direction of the anterior end of the cell — is induced by a solution containing 10 mM ATP, 10mM $CaCl_2$, and 0.1 mM $ZnCl_2$. Working with Paramecium models obtained by a new method, namely, extraction with Triton X-100, Naitoh (personal communication) showed that in a calcium-free solution (ATP + Mg + EGTA) these cell models swim forward, whereas Ca^{++} in a concentration of $1 \times 10^{-7} - 1 \times 10^{-6}$ g-ion liter in the presence of $(Mg-ATP)^{--}$ induces their movement backward.

To page 143

By the Method of Ulbrecht and Ulbrecht (1957). Material: muscles of a rabbit's feet and hind limbs.

The muscles are first put through a mincer and then homogenized for 2 min in a Waring blender. The homogenate is washed with 10 volumes 0.1 M KCl and centrifuged for 2 min at 800 g. The top layers of the residue are removed and further homogenization in the blender is carried out for 2 min in 0.1 M KCl; the resulting homogenate is twice washed with 15-20 volumes 0.1 M KCl, with centrifugation each time at 1800 g. The top layer of the last residue contains the myofibrils. This layer is removed and washed 7-10 times with 15-20 volumes 0.1 M KCl, and after each second washing the residue is homogenized in the blender for 30 sec. All procedures are carried out in the cold. Ultimately the top layers are collected and transferred to 50% glycerol at $-15°C$. The suspension contains 20-30 mg protein/ml and can be kept for many months without change in ATPase activity. Before the preparation is used the glycerol is removed by washing twice in 10-15 volumes 0.1 M KCl, and the residue after centrifugation is resuspended in 0.1 M KCl.

By the Method of Solaro et al. (1971). A preparation of pure myofibrils uncontaminated by mitochondrial membranes or membranes of the sarcolemma and sarcoplasmic reticulum. Material: hearts obtained from dogs anesthetized with pentobarbital. The heart is washed with 0.9% NaCl at 4°C. Fat and connective tissue are removed. The muscle is cut with scissors and then homogenized with 4 volumes 0.3 M sucrose with 10 mM imidazole, pH 7.0, for 1 min at 0.4°C. The homogenate is centrifuged for 20 min at 17,000-18,000 g. The residue is resuspended in standard salt solution (60 mM KCl, 30 mM imidazole, pH 7.0, 2 mM MgCl$_2$) equal in volume to the original homogenate. During this and subsequent resuspensions the bottom layer of the residue, consisting of coarse particles, is discarded. The suspension is again homogenized, as on the first occasion, and centrifuged for 15 min at 750 g. Resuspension, homogenization, and centrifugation are repeated four or more times. As a result, a light brown homogenate of washed myofibrils is obtained. They are resuspended in 8 volumes of standard salt solution also containing 1% Triton X-100. Centrifugation for 15 min at 750 g and homogenization follow. This and all subsequent homogenizations are carried out by hand in a glass homogenizer with teflon pestle. Treatment with Triton X-100 and homogenization are repeated once again. The residue is washed four times with 8 volumes standard salt solution to remove the Triton, and the gentle homogenization is then repeated. The completely purified myofibrils are resuspended in the standard salt solution to give a protein concentration of 10-15 mg/ml.

To page 145

(After Madeira and Carvalho, 1969.) Material: dorsal muscles of a rabbit. The muscles (35 g) are homogenized in 100 ml 50 mM CaCl$_2$ for 20 sec at 0-5°C (Philips model HM 3080 homogenizer). The homogenate is filtered through a nylon sieve with a pore area of 4 mm^2. The residue is resuspended in 100 ml CaCl$_2$ and rehomogenized repeatedly until the residue in the sieve is no longer brown. All the filtrates are pooled and again filtered. The suspension is homogenized at 3000 g for 5 sec. The residue is resuspended in 360 ml KCl-buffer: 50 mM KCl + 30 mM KHCO$_3$ + 2.5 mM histidine, pH 6, at 0-5°C. The suspension is centrifuged as before, and the residue is washed a further twice with KCl-buffer, each time in a volume of 360 ml. The residue is resuspended in 160 ml KCl-buffer in 4 tubes and continuously stirred for 30 min at 37°C. The tubes are then left to stand in ice for 10 min and a precipitate forms. It is resuspended in 360 ml KCl-buffer, allowed to stand for 10 min again in ice, and the supernatant is removed. Suspension and precipitation are repeated four times. The residue is then resuspended in 360 ml deionized water, and the suspension placed on ice for 10 min. The residue is suspended in 360 ml 2.5 × 10^{-7} M NaOH and kept on ice for 25 min. The residue swells considerably, so that it is possible to pour off only a small quantity of the supernatant. Suspension in 2.5 × 10^{-7} M NaOH and subsequent precipitation on ice are repeated until swelling of the precipitate ends, i.e., usually twice or three times. The precipitate is then resuspended once again in 2.5 × 10^{-7} M NaOH, but is then certrifuged at 3000 g for 15 sec. This procedure is repeated, but centrifugation lasts 5 min. The supernatant is poured off and the transparent part of the residue collected. The remaining residue is resuspended in the same NaOH solution and centrifuged as before. The transparent part of the residue is again collected and added to that obtained previously. To each 10 ml of residue 1 ml of 2 mM ATP, pH 7.0, is added and the suspension is mixed. After the addition of 25 volumes of deionized water at 0-5°C the suspension is centrifuged at 1500 g for 1 min. The residue thus obtained, the purity of which is tested under the phase-contrast microscope (200×), contains empty sarcolemmas.

To page 156

By the Method of Amos (1971). Vorticella convallaria is extracted for 1 month in the cold in ES: 50% glycerol, 0.1 M KCl, 10 mM histidine, and EGTA/Ca-EGTA buffer in a total concentration of 8 mM, pH 7.0. At [Ca^{++}] < 1 × 10^{-7} g-ion/liter the stalks are straightened and relaxed; at [Ca^{++}] > 5 × 10^{-7} g-ion/liter they are contracted and coiled. The threshold Ca^{++} concentration varies slightly if the preparation is kept in Ca^{++}-buffer overnight. By alternating solutions with 1 × 10^{-6} and 1 × 10^{-8} g-ion Ca^{++}/liter up to 35 contraction–relaxation cycles of the stalk could be obtained.

By the Method of Townes and Brown (1965). Cultures of Vorticella convallaria are kept in cerophyl–egg yolk fluid as described by Levine (J. Protozool., 6:163, 1959) on fragments of cover slips. The coverslips with the vorticellas are transferred 24 h before extraction into medium with cerophyl–aerobacter (Sonneborn, J. Exper. Zool., 113:87, 1950), for under these conditions the vorticellas develop longer stalks. The fragments of cover slips with the attached vorticellas are then transferred to cold ES: 50% glycerol, 4 mM EDTA, 5 mM histidine, pH 6.8. Extraction continues for 45 min at from -12 to -14°C, and then for 10 min at room temperature (24 ± 2°C), after which the residue is again washed for 5 min with 3 portions of ES, each of 1 ml, at the same room temperature (the total extraction time is thus 60 min). The reactivating solution contains 0.02 M maleate buffer, 0.12 M KCl, 3 × 10^{-9} g-ion Ca^{++}/liter, 4 × 10^{-3} g-ion Mg^{++}/liter, 6 × 10^{-3} M ATP. The stalks are completely relaxed in this medium at pH 6.8. Changing the pH to 7.0 by means of 0.1 N NaOH induces contraction of the stalks to 10% of their initial length. Returning the pH to 6.8 does not abolish this contraction, but treatment with EDTA solution leads to relaxation. Returning the stalks to the original RS leads to a fresh contraction. Alternation of contraction and relaxation could be repeated up to 8 times. Contraction can also be induced without changing the pH to 6.8 if Ca^{++} is added to the medium in a concentration of 3 × 10^{-6} g-ion/liter.

Bibliography

SURVEYS*

Aleksandrov, V. Ya., 1963, "Cytophysiological and cytoecological investigations of the resistance of plant cells to high and low temperatures," Trud. Bot. Inst. im. V. L. Komarova Akad. Nauk SSSR, Ser. 4, Eksper. Bot., 16:234. English translation (1964) in: Quart. Rev. Biol., 39(1):35.

Aleksandrov (Alexandrov), V. Ya., 1967, "Protein thermostability of a species and habitat temperature," in: Mechanisms of Temperature Adaptation (C. L. Prosser, ed.), Publication No 84 of the American Association for the Advancement of Science, Washington, D. C.

Aleksandrov (Alexandrov), V. Ya., 1969, "Conformational flexibility of proteins, their resistance to proteinases and temperature conditions of life," Currents in Modern Biology 3:9.

Aspects of Cell Motility. Symposia of the Society for Experimental Biology, Vol. 22, Cambridge, 1968.

Bendall, J. R., 1969, Muscles, Molecules, and Movement, Heinemann, London.

Bishop, D. W., 1958, "Sperm cell models and the question of ATP-induced rhythmic motility," Biol. Bull., 115(2):326.

Bishop, D. W., 1962, "Sperm motility," Physiol. Rev., 52(1):1.

Brokaw, C. J., 1966, "Mechanics and energetics of cilia," Amer. Rev. Resp. Dis., 93(3):32.

Brokaw, C. J., 1968, "Mechanisms of sperm movement," in: Symposia of the Society for Experimental Biology, Vol. 22, Cambridge, pp. 101-116.

Burnasheva, S. A., Fedorova, L. G., and Lyubimova, M. N., 1963, "Structural organization and biochemical basis of ciliary and flagellar movement," Uspekhi Sovr. Biol., 56(3):365.

Csapo, A., 1960, "Molecular structure and function of smooth muscle," in: The structure and Function of Muscle, Vol. 1, New York and London, pp. 229-264.

---

*Where these items are cited in the text the year of publication is underlined.

Doetsch, R. N., and Hageage, G. G., 1968, "Motility in procaryotic organisms: problems, points of view and perspectives," Biol. Rev., 43(3):317.

Ebashi, S., Endo, M., and Ohtsuki, I., 1969, "Control of muscle contraction," Quart. Rev. Biophys., 2(4):351.

Éngel'gardt, V. A., 1945, "Phosphoric acid and cell functions," Izvest. Akad. Nauk SSSR, Ser. Biol., 2:182.

Éngel'gardt (Engelhardt), V. A., 1946, "Adenosinetriphosphatase properties of myosin," Advances Enzymol., 6:147.

Éngel'gardt, V. A., 1957, "The chemical basis of the motile function of cells and tissues," Vestn. Akad. Nauk SSSR, 27(11):58.

Ernst, E., 1963, Biophysics of the Striated Muscle, Budapest.

Federation Proceedings, 1964, 23(5):885.

Federation Proceedings, 1965, 24(5):1112.

Gibbons, I. R., 1967, "The structure and comparison of cilia," in: Formation and Fate of Cell Organelles, New York and London, pp. 99-113.

Gray, J., 1955, "The movement of sea urchin spermatozoa," J. Exp. Biol., 32(4):775.

Hanson, J., and Huxley, H. E., 1955, "The structural basis of contraction in striated muscle," in: Symposia of the Society for Experimental Biology, Vol. 2, Cambridge, pp. 228-264.

Hanson, J., and Lowy, J., 1960, "Structure and function of the contractile apparatus in the muscles of invertebrate animals," in: The Structure and Functions of Muscle, Vol. 1, New York and London, 265-335.

Hasselbach, W., 1963, "Mechanismen der Muskelkontraktion und ihre Intrazelluläre Steuerung," Naturwissenschaften, 50(7):249.

Hasselbach, W., 1964, "Relaxing factor and the relaxation of muscle," Progress in Biophysics, 14:167.

Hasselbach, W., and Weber, A., 1955, "Models for the study of the contraction of muscle and of cell protoplasm," Pharmacol. Rev., 7(1):97.

Hoffmann-Berling, H., 1955, "Vergleich der Zellmotilität und der Muskelkontraktion," Rend. Ist. Lombardo Sci. e Lettere, 89:284.

Hoffmann-Berling, H., 1958, "Physiologie der Bewegungen und Teilungsbewegungen tierischer Zellen," Fortschr. Zool., 11:142.

Hoffmann-Berling, H., 1959, "The role of cell structures in cell movements," in: Cell, Organism and Milieu, New York, pp. 45-62.

Hoffmann-Berling, H., 1960, "Other mechanisms producing movements," in: Textbook of Comparative Biochemistry, Vol. 2, New York, pp. 341-370.

Hoffmann-Berling, H., 1963, "Mechanismen von Zellbewegungen," Naturwissenschaften, 50(7):256.

Huxley, H. E., 1965, "Structural evidence concerning the mechanism of contraction in striated muscle," in: Muscle, New York and Amsterdam, pp. 3-28.

Huxley, H. E., and Hanson, J., 1960, "The molecular basis of contraction of corssstriated muscles," in: The Structure and Function of Muscle, Vol. 1, New York and London, pp. 183-227.

Ivanov, I. I., 1968, "Biochemical mechanism of biphasic muscular activity," Uspekhi Sovr. Biol., 66(1):3.

Ivanov, I. I., and Yur'ev, V. A., 1961, The Biochemistry and Pathobiochemistry of Muscles [in Russian], Leningrad.

Kamiya, N., 1959, Protoplasmic Streaming, Protoplasmologia, Handbuch der Protoplasmaforschung, Vol. 8, Springer Verlag, Vienna.

Kamiya, N., 1968, "The mechanism of cytoplasmic movement in a myxomycete plasmodium," in: Symposia of the Society for Experimental Biology, Vol. 22, Cambridge, pp. 199-214.

Katz, B., 1966, Nerve, Muscle, and Synapse, McGraw-Hill, New York.

Kinosita, H., and Murakami, A., 1967, "Control of ciliary motion," Physiol. Rev., 47(1):53.

Krolenko, S. A., 1965, "Connection between bioelectrical phenomena and specific activity of cells," Tsitologiya, 7(4):480.

Kühne, W., 1864, Untersuchungen über das Protoplasma und die Kontraktilität, Leipzig.

Mazia, D., 1961, Mitosis and the Physiology of Cell Division, in: The Cell (J. Brachet and A. E. Mirsky, eds.), Vol. 3, New York.

Nasonov, D. N. and Aleksandrov, V. Ya., 1940, Reaction of Living Matter to the Action of External Agents [in Russian], Moscow—Leningrad.

Perry, S. V., 1955, "The components of the myofibril and their relation to its structure and function," in: Symposia of the Society for Experimental Biology, Vol. 9, Cambridge, pp. 203-264.

Perry, S. V., 1957, "Relationship between the chemical and contractile properties of skeletal muscle cells and their structure," in: Current Problems in Biochemistry [in Russian], Moscow, pp. 148-262.

Perry, S. V., 1967, "The structure and interaction of myosin," Progr. Biophys. Mol. Biol., 17:327.

Podolsky, R. J., 1968, "Membrane systems in muscle cells," in: Symposia of the Society for Experimental Biology, Vol. 22, Cambridge, pp. 87-99.

Poglazov, B. F., 1966, The Structure and Functions of Contractile Proteins, Academic Press, New York.

Pringle, J. W. C., 1967, "The contractile mechanism of insect fibrillar muscle," in: Progr. Biophys. Mol. Biol., 17(1):1.

Pringle, J. W. C., 1968, "Mechano-chemical transformation in striated muscle," Symposia of the Society for Experimental Biology, Vol. 22, Cambridge, pp. 67-86.

Rüegg, J. C., 1968, "Contractile mechanism of smooth muscle," in: Symposia of the Society for Experimental Biology, Vol. 22, Cambridge, pp. 45-66.

Seravin, L. N., 1967, Motile Systems of Protozoans [in Russian], Leningrad.

Szent-Györgyi, A., 1951, Chemistry of Muscular Contraction, New York.

Szent-Györgyi, A. G., 1960, "Proteins of the myofibril," in: Structure and Function of Muscle, Vol. 2, New York.

Ushakov, B. P., 1967, "Thermostability of cells and protoplasmic proteins in poikilothermic animals in relation to the problem of species," in: The Cell and Environmental Temperature, Proceedings of the International Symposium on Cytoecology (C. L. Prosser, ed.), Pergamon Press, Oxford, pp. 322-334.

Vorob'ev, V. I., 1966a, "Contractility," in: Textbook of Cytology [in Russian], Vol. 2, Moscow—Leningrad, pp. 67-91.

Vorob'ev, V. I., 1966b, "Contractile models and mechanochemistry," in: The Biophysics of Muscular Contraction [in Russian], Moscow, pp. 184-191.

Weber, H. H., 1951, "Die Aktomyosin modelle und der Kontraktions-zyklus des Muskels," Z. Elektrochem., 55(5):511.

Weber, H. H., 1953, "Muskelkontraktion, Zellmotilität und ATP," Biochim. Biophys. Acta, 12(1/2):150.
Weber, H. H., 1958, The Motility of Muscle and Cells, Cambridge.
Weber, H. H., 1960, "Der Mechanismus der Kontraktion und der Erschlaffung des Muskels," Arzneimittel-Forschung, 10(5):404.
Weber, H. H., and Portzehl, H., 1952a, "Kontraktion, ATP-Zyclus and fibrilläre Proteine des Muskels," Ergebn. Physiol., 47:369.
Weber, H. H., and Portzehl, H., 1952b, "Muscle contraction and fibrous muscle proteins," Advances Protein Chem., 7:161.
Wohlfarth-Bottermann, K. E., 1968, "Dynamik der Zelle," Mikroskopie, 23(1):71.
Zhukov, E. K., 1969, Essays on Neuromuscular Physiology [in Russian], Leningrad.

EXPERIMENTAL PAPERS

Abbott, R. H., and Chaplain, R. A., 1966, "Preparation and properties of the contractile element of insect fibrillar muscle," J. Cell Sci., 1(3):311.
Adelstein, R. S., Pollard, T. D. and Michael, W. M., 1971, "Isolation and characterization of myosin fragments from human blood platelets," Proc. Nat. Acad. Sci. USA, 68(11):2703.
Aleksandrov, V. Ya., 1952, "The correlation between the thermostability of protoplasm and the ambient temperature," Dokl. Akad. Nauk SSSR, 83(1):149.
Aleksandrov, V. Ya., 1954, "A simplified method of infiltrating plant tissues," Bot. Zh., 39 (2): 421
Aleksandrov, V. Ya., and Arronet, N. I., 1956, "Adenosinetriphosphate induces ciliary movement of the ciliated epithelium when killed by glycerol extraction ("cell model")," Dokl. Akad. Nauk SSSR, 110(3):457.
Aleksandrov, V. Ya., Arronet, N. I., Den'ko, E. I., and Konstantinova, M. F., (Alexandrov, V. Ja., Arronet, N. J., Denko, E. J., and Konstantinova, M. F.), 1965, "Influence of D_2O on resistance of plant and animal cells, cellular models and actomyosin to some denaturing agents," Nature, 205:286.
Aleksandrov, V. Ya., Arronet, N. I., Den'ko, E. I., and Konstantinova, M. F. (Alexandrov, V. Ja., Arronet, N. J., Denko, E. J., and Konstantinova, M. F.), 1966, "Effect of heavy water (D_2O) on resistance of plant and animal cells, cell models and protein to certain denaturing factors," Federat. Proc., 25(1):128.
Aleksandrov, V. Ya., and Vol'fenzon, L. G., 1956, "Reversible contractions of connective-tissue cells upon the action of various agents," Zh. Obshch. Biol., 17(2):142.
Amos, W. B., 1971, "Reversible mechanochemical cycle in the contraction of Vorticella," Nature, 229(5280):127.
Aronson, J. F., 1965, "The use of fluorescein-labeled heavy meromyosin for the cytological demonstration of actin," J. Cell Biol., 26(1):293.
Arronet, N. I., 1964a, "The injurious action of a high hydrostatic pressure and of heat on ciliated epithelial cells and the portective effect of glycerol," Tsitologiya, 6(4):432.
Arronet, N. I., 1964b, "The point of application of the injurious action of heat on the cell," Dokl. Akad. Nauk SSSR, 157(2):437.

BIBLIOGRAPHY

Arronet, N. I., 1968, "The action of certain metabolic inhibitors depressing motility of ciliated epithelial cells," Dokl. Akad. Nauk SSSR, 179(4):217.

Arronet, N. I., and Konstantinova, M. F., 1964, "The point of application of the injurious effect of a high hydrostatic pressure on the cell," Tsitologiya, 6(6):743.

Arronet, N. I., and Konstantinova, M. F., 1969, "The harmful action of a high hydrostatic pressure on muscle," Tsitologiya, 11(2):148.

Ashhurst, D. E., 1969, "The effect of glycerination on the fibrillar flight muscles of Belostomatid water-bugs," Z. Zellforsch., 93(1):36.

Ashley, C. C., and Ridgway, E. B., 1968, "Simultaneous recording of membrane potential, calcium transient and tension in single muscle fibers," Nature, 219:1168.

Bailey, K., 1956, "Invertebrate tropomyosin," Biochem. J., 64:9P.

Beck, R., Komnick, H., Stockem, W., and Wohlrafth-Bottermann, K. E., 1969, "Weitreichende fibrilläre Protoplasmadifferenziehrungen und ihre Bedeutung für die Protoplasmabewegung. IV. Vergleichende Untersuchungen an Actomyosin-Fäden und glycerinierten Zellen," Cytobiologie, 1(1):99.

Behnke, O., Kristensen, B. I., and Nielsen, L. E., 1971a, "Electron microscopical observations on actinoid and myosinoid filaments in blood platelets," J. Ultrastruct. Res., 37(3-4):351.

Behnke, O., Forer, A., and Emmerson, J., 1971b, "Actin in sperm tails and meiotic spindles," Nature, 234(5329):408.

Bemis, J. A., Bryant, G. M., Arcos, J. C., and Argus, M. F., 1968, "Swelling and contraction of mitochondrial particles: a re-examination of the existence of a contractile protein extractable with 0.6 M potassium chloride," J. Mol. Biol., 33(1):299.

Bendall, J. R., 1952, "A factor modifying the shortening response of muscle fibre bundles to ATP," Proc. Roy. Soc., B, 139:908.

Bendall, J. R., 1954, "The relaxing effect of myokinase on muscle fibres: its density with the "Marsh" factor," Proc. Roy. Soc., B, 142:409.

Bettex-Galland, M., and Lüscher, E. F., 1961, "Thrombosthenin — a contractile protein from thrombocytes. The extraction from human blood platelets and some of its properties," Biochim. Biophys. Acta, 49(3):536.

Bettex-Galland, M., Portzehl, H., and Lüscher, E. F., 1962, "Dissociation of thrombosthenin into two components comparable with actin and myosin," Nature, 193:777.

Beyersdorfer, K., 1951, "Gegenseitige Anziehung gleicher Strukturelemente bei Kollagen-Fibrillen und Trichocysten," Z. Naturforsch., 6b(2):57.

Bishop, D. W., 1958a, "Sperm contractile protein," Anat. Rec., 131(3):533.

Bishop, D. W., 1958b, "Glycerine-extracted models of sperm of the squid Loligo pealii," Anat. Rec., 132(3):414.

Bishop, D. W., 1958c, "Relaxing factors in ATP-induced motility of sperm models," Anat. Rec., 132(3):414.

Bishop, D. W., 1958d, "Mammalian sperm cell models reactivated by ATP," Federat. Proc., 17(1):15.

Bishop, D. W., and Hoffmann-Berling, H., 1959, "Extracted mammalian sperm models," J. Cell. Comp. Physiol., 53(3):445.

Bohr, D. F., Filo, R. S., and Gathe, K. F., 1962, "Contractile protein in vascular smooth muscle," Physiol. Rev., 42, Suppl. 5:98.

Borbire, M., and Szent-Györgyi, A. G., 1949, "On the relation between tension and ATP in cross-striated muscle," Biol. Bull., 96(2):162.

Bozler, E., 1956, "The effect of polyphosphates and magnesium on the mechanical properties of extracted muscle fibers," J. Gen. Physiol., 39(5):789.

Brokaw, C. J., 1960, "Decreased adenosine triphosphatase activity of flagella from a paralyzed mutant of Chlamydomonas moewusii," Exp. Cell Res., 19(2):430.

Brokaw, C. J., 1961, "Movement and nucleoside polyphosphatase activity of isolated flagella from Polytoma uvella," Exp. Cell. Res., 22(1):151.

Brokaw, C. J., 1963, "Movement of the flagella of Polytoma uvella," J. Exp. Biol., 40(1):149.

Brokaw, C. J., 1967, "Adenosine triphosphate usage by flagella," Science, 156:76.

Buchtal, F., Deutsch, A., Knappeis, G. G., and Munch-Peterson, A., 1946, "On the effect of adenosinetriphosphate on myosin threads," Acta Physiol. Scand., 13:167.

Burnasheva, S. A., Efremenko, M. V., and Lyubimova, M. N., 1963, "Investigation of adenosine-triphosphatase activity of isolated cilia of the infusorian Tetrahymena pyriformis and isolation of adenosine triphosphatase from them," Biokhimiya, 28(3):541.

Burnasheva, S. A., Efremenko, M. V., and Chumakova, L. P., 1964, "Structural organization and some biochemical bases of ciliary movement," Abstracts of Proceedings of the First All-Union Biochemical Congress [in Russian], Vol. 1, Moscow—Leningrad, pp. 46-47.

Burnasheva, S. A., Yurzina, G. A., and Lyubimova, M. N., 1969, "Ultrastructure of the flagella of Stigomonas oncopelti and the cilia of Tetrahymena pyriformis and the localization of adenosine triphosphatase (ATPase) in them," Tsitologiya, 11(6):695.

Carasso, N., and Favard, P., 1966, "Mise en évidence du calcium dans les myonèmes péndoculaires," J. Microscopie, 5(6):759.

Chaplain, R. A., 1966, "Tension development of glycerinated insect muscle fibres as a measure of the conformational state of the myosin," Biochem. Biophys. Res. Commun., 24(4):526.

Chaplain, R. A., 1970, "On the role of troponin in the mechanism of muscular contraction," Acta Biol. Med. Germ., 24(4):483.

Chernokozheva, I. S., 1967a, "A study of the thermostability of muscles and muscle models in connection with the growth of frogs," in: Variation in Thermostability of Animal Cells in Ontogenesis and Phylogenesis [in Russian], Leningrad, pp. 13-19.

Chernokozheva, I. S., 1967b, "A study of the thermostability of muscles and muscle models of juvenile frogs in the early seasons of the year," in: Variation in Thermostability of Animal Cells in Ontogenesis and Phylogenesis [in Russian], Leningrad, pp. 20-26.

Child, F. M., and Tamm, S., 1963, "Metachronal ciliary coordination in ATP-reactivated models of Modiolus gills," Biol. Bull., 125(2):373.

Costantin, L. L., and Podolsky, R. J., 1967, "Depolarization of the internal membrane system in the activation of frog skeletal muscle," J. Gen. Physiol., 50(5):1101.

Csapo, A., 1949, "Studies on adenosine triphosphatase activity of uterine muscle," Acta Physiol. Scand., 19(1):100.

BIBLIOGRAPHY

Csapo, A., 1959, "Studies on excitation—contraction coupling," Ann. New York Acad. Sci., 81:453.

Dörr, D., and Portzehl, H., 1954, "Die kontraktile Myosinfäden aus glatter Muskulatur," Z. Naturforsch., 9b(8):550.

Dregol'skaya, I. N., 1963, "Thermostability of the ciliated epithelium of the gills of the Black Sea mussel at different seasons of the year," in: Problems in the Cytoecology of Animals [in Russian], Moscow—Leningrad, pp. 43-50.

Dregol'skaya, I. N., and Chernokozheva, I. S., 1970, "The thermostability of muscles, of ciliated epithelial cells, and of their glycerinated models obtained from Rana temporaria at different seasons of the year," Tsitologiya, 12(1):51.

Eisenberg, B., and Eisenberg, R. S., 1968, "Selective disruption of the sarcotubular system in frog skeletal muscle," J. Cell Biol., 39(2):451.

Eisenberg, R. S., and Gage, R. W., 1967, "Frog skeletal muscle fibers: changes in electrical properties after disruption of transverse tubular system," Science, 158:1700.

El'tsina, N. V., 1948, "Adenosine-triphosphatase activity of structural proteins," Biokhimiya, 13(2):351.

Engel'gardt, V. A., and Burnasheva, S. A., 1957, "Localization of spermazin protein in spermatocytes," Biokhimiya, 22(3):554.

Engel'gardt, V. A., and Lyubimova, M. N. (Engelhardt, V. A., and Ljubimova, M. N.), 1939, "Myosin and adenosine triphosphatase," Nature, 144:668.

Engel'gardt, V. A., and Lyubimova, M. N., 1942, "The mechano-chemistry of muscle," Biokhimiya, 7(5-6):205.

Engel'gardt, V. A., Lyubimova, M. N., and Meitina, R. A., 1941, "Chemical and mechanical properties of muscles as shown by experiments on the myosin thread," Dokl. Akad. Nauk SSSR, 30(7):639.

Favard, P., and Carasso, N., 1965, "Mise en évidence d'un reticulum endoplasmique dans le spasmonème de ciliés Péritriches," J. Microscopie, 4:567.

Filo, R. S., Bohr, D. F., and Rüegg, J. C., 1965, "Glycerinated skeletal and smooth muscle: calcium and magnesium dependence," Science, 147:1581.

Gage, R. W., and Eisenberg, R. S., 1967, "Action potentials without contraction in frog skeletal muscle fibers with disrupted transverse tubules," Science, 158:1702.

Ganelina, L. Sh., 1962, "Effect of stretching muscle models on their resistance to denaturing agents," Tsitologiya, 4(2):223.

Gawadi, N. 1971, "Actin in the mitotic spindle," Nature, 234(5329):410.

Gibbons, I. R., 1965, "Reactivation of glycerinated cilia from Tetrahymena pyriformis," J. Cell Biol., 25(2):400.

Gibbons, I. R., and Rowe, A. J., 1965, Dynein: a protein with adenosine triphosphatase activity from cilia," Science, 149:424.

Gicquaud, C. R., and Couillard, P., 1970, "Préservation des mouvements dans le cytoplasme démembrané d'Amoeba proteus," Cytobiologie, 1(4):460.

Goodall, M. C., 1956, "Auto-oscillations in extracted muscle systems," Nature, 177:1238.

Goodall, M. C., and Szent-Györgyi, A. G., 1953, "Relaxing factor in muscle," Nature, 172:84.

Hanson, J., 1952, "Changes in the cross-striation of myofibrils during contraction induced by adenosine triphosphate," Nature, 169:530.

Hanson, J., and Lowy, J., 1963, "The structure of F-actin and actin filaments isolated from muscle," J. Mol. Biol., 6(1):46.

Hasselbach, W., 1952, "Die Diffusionskonstante des Adenosin-triphosphats in Innern der Muskulatur," Z. Naturforsch., 7a(6):334.

Hatano, S., and Oosava, F., 1966, "Extraction of an actin-like protein from the plasmodium of a myxomycete and its interaction with myosin A from rabbit striated muscle," J. Cell Physiol., 68(2):197.

Hatano, S., and Tazava, M., 1968, "Isolation, purification and characterisation of myosin B from myxomycete plasmodium," Biochim. Biophys. Acta, 154(3):507.

Hayashi, T., 1951, "Contractile properties of compressed monolayers of actomyosin," Federat. Proc., 10(1):61.

Hayashi, T., 1952, "Contractile properties of compressed monolayers of actomyosin," J. Gen. Physiol., 36(2):139.

Hoffmann-Berling, H., 1953, "Die wasser-glyzerin-extrahierte Zelle als Modell der Zellmotilität," Biochim. Biophys. Acta, 10(4):628.

Hoffmann-Berling, H., 1954a, "Adenosintriphosphate als Betriebsstoff von Zellbewegungen," Biochim. Biophys. Acta, 14(2):182.

Hoffmann-Berling, H., 1954b, "Die Bedeutung des Adenosintriphosphat für die Zell- und Kernteilungsbewegungen in der Anaphase," Biochim. Biophys. Acta, 15(2):226.

Hoffmann-Berling, H., 1954c, "Die glyzerin-wasserextrahierte Telophasezelle als Modell der Zytokinese," Biochim. Biophys. Acta, 15(3):332.

Hoffmann-Berling, H., 1955, "Geisselmodelle und Adenosintriphosphat," Biochim. Biophys. Acta, 16(1):146.

Hoffmann-Berling, H., 1956, "Das kontraktyle Eiweiss undifferenzierter Zellen," Biochim. Biophys. Acta, 19(3):453.

Hoffmann-Berling, H., 1958, "Der Mechanismus eines neuen, von der Muskelkontraktion verschiedenen Kontraktions-zyklus," Biochim. Biophys. Acta, 27(2):247.

Hoffmann-Berling, H., 1964, "Relaxation of fibroblast cells," in: Primitive motile systems in cell biology, New York, pp. 365-375.

Hoffmann-Berling, H., and Weber, H. H., 1953, "Vergleich der Motilität von Zellmodellen und Muskelmodellen," Biochim. Biophys. Acta, 10(4):629.

Huxley, H. E., 1963, "Electron microscope studies on the structure of striated muscle," J. Mol. Biol., 7(3):281.

Huxley, H. E., 1968, "Structural difference between resting and rigor muscle; evidence from intensity changes in the low-angle equatorial X-ray diagram," J. Mol. Biol., 37(3):507.

Huxley, H. E., 1969, "The mechanism of muscular contraction," Science, 164:1356.

Huxley, H. E., and Brown, W., 1967, "The low-angle X-ray diagram of vertebrate striated muscle and its behaviour during contraction and rigor," J. Mol. Biol., 30(2):383.

Huxley, H. E., and Hanson, J., 1954, "Changes in the cross-striations of muscle during contraction and stretch and their structural interpretation," Nature, 173:973.

Ishikawa, H., Bishoff, R., and Holzer, H., 1969, "Formation of arrowhead complexes with heavy meromyosin in a variety of cell types," J. Cell. Biol., 43:312.

BIBLIOGRAPHY

Ivanov, I. I., Gaitskhoki, V. S. and Korkhov, V. V., 1959, "The action of x-rays on motor function and contractile proteins of motile cells," Byull. Éksperim. Biol. i Med., (12):47.

Ivanov, I. I., and Torchinskii, Yu. M., 1955, "The nature of contraction of actomyosin and monolayer actomyosin filaments under the influence of adenosine triphosphate," Biokhimiya, 20(2):328.

Jagendorf-Elfvin, M., 1967, "Ultrastructure of the contraction-relaxation cycle of glycerinated rabbit psoas muscle. II. The ultrastructure of glycerinated fibers relaxed in EDTA and ATP following ATP-induced contraction," J. Ultrastruct. Res., 17(3-4):379.

Jewell, B. R., Pringle, J. W. S., and Rüegg, J. C., 1964, "Oscillatory contraction of insect fibrillar muscle after glycerol extraction," J. Physiol. (London), 173(2):6P.

Jewell, B. R., and Rüegg, J. C., 1966, "Oscillatory contraction of insect fibrillar muscle after glycerol extraction," Proc. Roy. Soc. B, 164:428.

Josephs, R., and Harrington, F., 1966, "Studies on the formation and physical chemical properties of synthetic myosin filaments," Biochemistry, 5:3474.

Kalamkarova, M. B., 1966, "The action of inhibitors of adenosine triphosphatase and cholinesterase on the mechanical properties of glycerinated muscle fibers," in: The Biophysics of Muscle Contraction [in Russian], Moscow, pp. 175-180.

Kalamkarova, M. B., and Veretennikova, A. A., 1970, "Changes of enzymatic properties of the muscle fibers during glycerol extraction, " Biofizika, 15 (1): 184.

Kaminer, B., 1960, "Effect of heavy water on different types of muscle and on glycerol extracted psoas fibres," Nature, 185:172.

Kaminer, B., 1966, "Effect of deuterium oxide on the relaxation of glycerol-extracted muscle," Nature, 209:809.

Kaminer, B., and Bell, A., 1966, "Myosin filamentogenesis: effect of pH and ionic concentration," J. Mol. Biol., 20(2):391.

Kamitsubo, E., 1969, "Motile protoplasmic fibrils in cells of Characeae," J. Cell Biol., 42:166a.

Kamiya, N., and Kuroda, K., 1965, "Movement of the myxomycete plasmodium. I. A study of glycerinated models," Proc. Japan Acad., 41(9):837.

Kazakova, T. B., 1964, "Contractile properties of glycerinated models of liver mitochondria," Biokhimiya, 29(1):35.

Kazakova, T. B., and Neifakh, S. A., 1963, "Mechanochemical activity and membrane permeability of mitochondria of normal and tumor cells," Dokl. Akad. Nauk SSSR, 152(2):471.

Keyserlingk, D., 1968, "Elektronenmikroskopische Untersuchung über die Differenzierungsvorgänge in Cytoplasma von segmentierten neutrophilen Leukozyten während der Zellbewegung, " Exp. Cell Res., 51(1):79.

Keyserlingk, D., and Schwarz, W., 1968, "Feinstruktur des kontraktilen Systems in Fibroblasten," Naturwissenschaften, 55(11):549.

Kinoshita, S., 1958, "The mode of action of metal-chelating substances on sperm motility in some marine forms as shown by glycerol-extracted sperm-models," J. Fac. Sci. Univ. Tokyo, Sec. 4, 8(2):219.

Kinoshita, S., 1959, "On the identity of the motility-inducing factor of flagellum and the relaxing factor of muscle," J. Fac. Sci. Univ. Tokyo, Sec. 4, 8(3):427.

Kinoshita, S., 1968, "Relative deficiency of intracellular relaxing system observed in presumptive furrowing region in induced cleavage in the centrifuged sea urchin egg," Exp. Cell. Res., 51(2):395.

Kinoshita, S., 1969, "Periodical release of heparin-like polysaccharide with cytoplasm during cleavage of sea-urchin egg," Exp. Cell. Res., 56(1):39.

Kinoshita, S., Andoh, B., and Hoffmann-Berling, H., 1964, "Das Erschlaffungssystem von Fibroblastenzellen," Biochim. Biophys. Acta, 79(1):88.

Kinoshita, S., and Hoffmann-Berling, H., 1964, "Lokale Kontraktion als Ursache der Plasmateilung von Fibroblasten," Biochim. Biophys. Acta, 79(1):98.

Kinoshita, S., and Yazaki, J., 1967, "The behaviour and localization of intracellular relaxing system during cleavage in the sea urchin egg," Exp. Cell. Res., 47(2):449.

Komissarchik, Ya. Yu., and Shapiro, E. A., 1969, "Ultrastructural organization of glycerinated frog fibers under normal conditions and after exposure to certain factors," Proceedings of the Second All-Union Conference on Electron Microscopy [in Russian], Kiev, pp. 37-38.

Komissarchik, Ya. Yu., and Shapiro, E. A., 1971, "Electron-microscopic investigation of glycerinated muscle fibers," Tsitologiya, 13:4.

Kono, T., and Colowick, S. P., 1961, "Isolation of skeletal muscle cell membrane and its properties," Arch. Biochem. Biophys., 93(3):520.

Kono, J., Kakuma, F., Homma, M., and Fukuda, S., 1964, "The electron-microscopic structure and chemical composition in the isolated sarcolemma of the rat skeletal muscle cell," Biochim., Biophys. Acta, 88(1):155.

Korey, S., 1950, "Some factors influencing the contractility of a non-conducting fiber preparation," Biochim. Biophys. Acta, 4(1-3):58.

Krolenko, S. A., 1967, "Vacuolation of the T-system of frog cross-striated muscle fibers," Proceedings of a Scientific Conference of the Institute of Cytology, Academy of Sciences of the USSR, to Commemorate the 50th Anniversary of the Great October Socialist Revolution [in Russian], Leningrad, pp. 57-58.

Krüger, F., Wohlfarth-Bottermann, K. E., and Pfefferkorn, G., 1952, "Die Trichocysten von Uronema marinum Dujardin," Z. Naturforsch., 7b(7):407.

Laki, K., 1967, "The contraction of glycerol-treated fibers in Thulet's solution," Biochem. Biophys. Res. Commun., 28(1):103.

Laki, K., and Bowen, W. J., 1955, "The contraction of muscle fiber and myosin B thread in KI and KCNS solutions," Biochim. Biophys. Acta, 16(2):301.

Lee, K. S., 1961, "Effect of ouabain on glycerol extracted fibers from heart containing a 'relaxing factor,'" J. Pharmacol. Exp. Ther., 132(1):149.

Lehninger, A. L., 1959a, "Reversal of thyroxine-induced swelling of rat liver mitochondria by adenosine triphosphate," J. Biol. Chem., 234(8):2187.

Lehninger, A. L., 1959b, "Reversal of various types of mitochondrial swelling by adenosine triphosphate," J. Biol. Chem., 234(9):2465.

Levin, S. V., Shapiro, E. A., and Shishina, N. N., 1967, "Binding of neutral red during contraction of glycerinated muscle fibers," Tsitologiya, 9(10):1281.

Levine, L., 1956, "Contractility of glycerinated Vorticellae," Biol. Bull., 111(2):319.

Loewy, A. G., 1951, "An actomyosin-like substance from the plasmodium of a myxomycete," J. Cell Comp. Physiol., 40(1):127.
Lorand, L., 1953, "Adenosine triphosphate−creatine transphorylase as relaxing factor of muscle," Nature, 172:1181.
Lorand, L., and Moos, C., 1956, "Auto-oscillations in extracted muscle systems," Nature, 177:1239.
Lowy, J., and Hanson, J., 1962, "Ultrastructure of invertebrate smooth muscles," Physiol. Rev., 42, Suppl. 5:34.
Lyubimova, M. N., Demyanovskaya, N. S., and Fain, F. S., 1964, "Connection between nucleotidases (di and triphosphatases) of higher plants with movement and growth. ATPase and ADPase," Abstracts of Proceedings of the First All-Union Biochemical Congress [in Russian], Vol. 1, Moscow−Leningrad, p. 46.
Lyubimova, M. N., Demyanovskaya, N. S., and Fain, F. S., 1966, "The mechanochemistry of leaf movement of the higher plant Mimosa pudica," Abstracts of Proceedings of an All-Union Conference on Muscle Biochemistry [in Russian], Moscow−Leningrad, p. 79.
Lyubimova, M. N., and Engel'gardt, V. A., 1939, "Adenosine triphosphatase and myosin of muscle," Biokhimiya, 4(6):716.
Lyudkovskaya, R. G., and Kalamkarova, M. B., 1966, "On the mechanism of isometric contraction of muscle," in: The Biophysics of Muscular Contraction [in Russian], Moscow, pp. 162-167.
Madeira, V. M. C., and Carvalho, A. P., 1969, "Interaction of ATP, ADP, and AMP with sarcolemma isolated from rabbit skeletal muscle," Mem. e Estud. Mus. Zool. Univ. Coimbra, 310:1.
Makhlin, E. E., and Scholl, E. D., 1968, "Effect of keeping temperature on thermostability of muscle tissues, muscle models, and aldolase of Barents Sea mollusks," Tsitologiya, 10(11):1442.
Marsh, B. B., 1951, "A factor modifying muscle fiber synaeresis," Nature, 167:1065.
Marsh, B. B., 1952, "The effects of adenosine triphosphate on the fiber volume of a muscle homogenate," Biochim. Biophys. Acta, 9(3):247.
Mazia, D., Petzelt, C., Williams, R. O., and Mazia, I, 1972, "A Ca-activated ATPase in the mitotic apparatus of the sea urchin egg (isolated by a new method)," Exper. Cell Res., 70(2):325.
Morgan, J., Fyfe, D., and Wolpert, L., 1967, "Isolation of microfilaments from Amoeba proteus," Exp. Cell Res., 48(1):194.
Nachmias, V. T., and Huxley, H. E., 1970, "Electron microscope observations on actomyosin and actin preparations from Physarum polycephalum, and on their interaction with heavy meromyosin subfragment I from muscle myosin," J. Mol. Biol., 50(1):83.
Naitoh, Y., 1966, "Reversal response elicited in nonbeating cilia of Paramecium by membrane depolarization," Science, 154:660.
Naitoh, Y., 1969, "Control of the orientation of cilia by adenosine triphosphate, calcium and zinc in glycerol-extracted Paramecium caudatum," J. Gen. Physiol., 53(5):517.
Naitoh, Y., and Jasumasu, I., 1967, "Binding of Ca-ions by Paramecium," J. Gen. Physiol., 50(5):1303.

Nakajima, H., 1960, "Some properties of a contractile protein in a myxomycete plasmodium," Protoplasma, 52:413.
Nakazawa, T., 1964, "Contraction of glycerinated mitochondria induced by ATP and divalent cations," J. Biochem., 56(1):22.
Nayler, W. G., and Merrilles, N. C. R., 1964, "Some observations on the fine structure and metabolic activity of normal and glycerinated ventricular muscle of toad," J. Cell Biol., 22(3):533.
Needham, D. M., 1950, "Myosin and adenosine triphosphate in relation to muscle contraction," Biochim. Biophys. Acta, 4(1-3):42.
Needham, D. M., 1962, "Contractile proteins in smooth muscle of the uterus," Physiol. Rev., 42, Suppl. 5:88.
Needham, J., Shin-Chang-Shen, Needham, D. M., and Lawrence, A. C., 1941, "Myosin birefringence and adenylpyrophosphate," Nature, 147:766.
Nelson, L., 1966, "Contractile proteins of marine invertebrate spermatozoa," Biol. Bull., 130(3):378.
Ohnishi, T., 1962, "Extraction of actin- and myosin-like proteins from erythrocyte membrane," J. Biochem., 52(4):307.
Ohnishi, Ts., and Ohnishi, T., 1962a, "Extraction of contractile protein from liver mitochondria," J. Biochem., 51(4):380.
Ohnishi, Ts., and Ohnishi, T., 1962b, "Extraction of actin- and myosin-like proteins from liver mitochondria," J. Biochem., 52(3):230.
Ovsyanko, E. P., 1968, "Sorption of basic and acid dyes by glycerinated muscle fibers upon contraction induced by ATP or heating," Tsitologiya, 10(7):844.
Packer, L., and Marchant, R., 1964, "Action of adenosine triphosphate on chloroplast structure," J. Biol. Chem., 239(6):2061.
Packer, L., and Young, J. A., 1965, "Retention of energy transfer in glycerinated chloroplasts," Biochem. Biophys. Res. Commun., 19(5):671.
Perry, S. V., 1951, "The adenosine triphosphatase activity of myofibrils isolated from skeletal muscle," Biochem. J., 48(3):257.
Perry, S. V., and Horne, R. W., 1952, "The intracellular components of skeletal muscle," Biochim. Biophys. Acta, 8(5):483.
Petzelt, C., 1972, "Ca^{++}-activated ATPase during the cell cycle of the sea urchin Strongylocentrotus purpuratus," Exper. Cell Res., 70(2):333.
Poglazov, B. F., 1956, "Adenosinetriphosphatase activity and the motile response in plants," Dokl. Akad. Nauk SSSR, 109(3):597.
Poglazov, B. F., 1964, "Contractile systems of living organisms," Abstracts of Proceedings of the First All-Union Biochemical Congress [in Russian], Vol. 1, Moscow—Leningrad, p. 21.
Pollard, T. D., Shelton, E., Weihing, R. R., and Korn, E. D., 1970, "Ultrastructural characterization of F-actin isolated from Acanthamoeba casellanii and identification of cytoplasmic filaments as F-actin by reaction with rabbit heavy meromyosin," J. Mol. Biol., 50(1):91.
Portzehl, H., 1951, " Muskelkontraktion und Modelkontraktion," Z. Naturforsch., 6b(7):855
Portzehl, H., 1952, "Der Arbeitszyklus geordneter Aktomyosinsystems (Muskel und Muskelmodelle)," Z. Naturforsch., 7b(1):1.
Portzehl, H., 1954, "Gemeinsame Eigenschaften von Zell- und Muskelkontraktilität," Biochim. Biophys. Acta, 14(2):195.

Portzehl, H., 1957, "Die Bindung des Erschlaffungsfaktors von Marsh an die Muskelgrana," Biochim. Biophys. Acta, 26(2):373.
Portzehl, H., Caltwell, P. C., and Rüegg, J. C., 1964, "The dependence of contraction and relaxation of muscle fibres from crab Maja squinado on the internal concentration of the free calcium ions," Biochim. Biophys. Acta, 79(3):581.
Portzehl, H., Schramm, G., and Weber, H. H., 1950, "Aktomyosin und seine Komponenten," Z. Naturforsch., 5b(1):61.
Raff, E. C., and Blum, J. J., 1966, "The effects of adenosine triphosphate and related compounds on some hydrodynamic properties of glycerinated cilia," J. Cell Biol., 31(3):445.
Ramsey, R. W., and Street, S. F., 1940, "The isometric length–tension diagram of isolated skeletal muscle fibres," J. Cell. Comp. Physiol., 15(1):11.
Randall, I. T., 1957, "Observation on contractile systems," J. Cell. Comp. Physiol., 49, Suppl. 1:199.
Ranney, R. C., 1954, "Spontaneous relaxation in glycerol-extracted muscle fiber bundles," Amer. J. Physiol., 179(1):99.
Reedy, M. K., Holmes, K. C., and Tregear, R. T., 1966, "Induced changes in orientation of the cross-bridges of glycerinated insect flight muscle," Nature, 207:1276.
Renaud, F. L., Rowe, A. J., and Gibbons, I. R., 1968, "Some properties of the protein forming the outer fibers of cilia," J. Cell Biol., 36(1):79.
Ridgway, E. B., and Ashley, C. C., 1967, "Calcium transients in single muscle fibers," Biochem. Biophys. Res. Commun., 29(2):229.
Rosental, S. L., Edelman, P. M., and Schwartz, J. L., 1965, "A method for the preparation of skeletal muscle sarcolemma," Biochim. Biophys. Acta, 109(2):512.
Rüegg, J. C., 1961a, "The proteins associated with contraction in lamellibranch "catch" muscle," Proc. Roy. Soc. B, 154:209.
Rüegg, J. C., 1961b, "On the tropomyosin-paramyosin system in relation to the viscous tone of lamellibranch "catch" muscle," Proc. Roy. Soc. B, 154:224.
Satir, P., and Child, F. M., 1963, "The microscopy of ATP-reactivated ciliary models," Biol. Bull., 125(2):390.
Schädler, M., 1967, "Proportionale Aktivierung von ATPase-Aktivität und Kontraktionsspannung durch Galciumionen in isolierten kontraktilen Strukturen verschiedener Muskularten," Pflügers Arch. Ges. Physiol., 296(1):70.
Schäfer-Danneel, S., 1967, "Strukturelle und funktionelle Voraussetzungen für die Bewegung von Amoeba proteus," Z. Zellforsch., 78(4):441.
Schäfer-Danneel, S., and Weissenfels, N., 1969, "Licht- und elektronenmikroskopische Untersuchungen über die ATP-abhängige Kontraktion kultivierter Fibroblasten nach Glycerin-Extraktion," Cytobiologie, 1(1):85.
Schick, A. F., and Hass, G. M., 1949, "A new method for isolation and purification of mammalian striated myofibrils," Science, 109:486.
Seidel, J. C., and Gergely, J., 1963, "Studies on myofibrillar adenosine triphosphatase with calcium-free adenosine triphosphate," J. Biol. Chem., 238(11):3648.
Senda, N, Shibata, N., Tatsumi, Noriyuki, Kondo, K., and Hamada, K., 1969, "A contractile protein from leucocytes. Its extraction and some of its properties," Biochim. Biophys. Acta, 181(1):191.
Seravin, L. N., 1961, "The role of adenosine triphosphate in beating of infusorian cilia," Biokhimiya, 26(1):160.

Seravin, L. N., 1963, "New methods of preparing contractile models from the Vorticella stalk," Biokhimiya, 28(4):606.

Seravin, L. N., Skoblo, I. I., and Bagnyuk, I. G., 1965, "Mechanism of contraction of myonemes in the infusorian Spirostomum ambiguum," Acta Protozool., 3(30): 329.

Shtrankfel'd, I. G., Filatova, L. G., and Kalamkarova, M. B., 1966, "Optical and mechanical properties of glycerinated muscle fibers after selective extraction of myosin and actin," in: The Biophysics of Muscle Contraction [in Russian], Moscow, pp. 180-184.

Simard-Duquesne, N., and Couillard, P., 1962a, "Ameboid movement. I. Reactivation of glycerinated models of Amoeba proteus with adenosine triphosphate," Exp. Cell Res., 28(1):85.

Simard-Duquesne, N., and Couillard, P., 1962b, "Ameboid movement. II. Research of contractile proteins in Amoeba proteus," Exp. Cell Res., 28(1):92.

Sjöstrand, F. S., and Jagendorf-Elfvin, M., 1967, "Ultrastructure studies of the contraction—relaxation cycle of glycerinated rabbit muscle. I. The ultrastructure of glycerinated fibers contracted by treatment with ATP," J. Ultrastruct. Res., 17(3-4):348.

Skholl', E. D., 1970, "Selection of a common vole population by thermostability of their cells," in: Proceedings of a Scientific Conference of the Institute of Cytology, Academy of Sciences of the USSR [in Russian], Leningrad, pp. 82-83.

Solaro, R. J., Pang, D. C. and Briggs, F. N., 1971, "The purification of cardiac myofibrils with triton X-100," Biochim. Biophys. Acta, 245(1):259.

Ströbel, G., 1952, "Doppelnrechungsänderungen bei der aktiven Kontraktion des Fasermodells (aus Kaninchenpsoas)," Z. Naturforsch., 7a(2):102.

Suzdal'skaya, I. P., 1964, "Vital staining of muscle tissue during excitation," Tsitologiya, 6(2):196.

Suzdal'skaya, I. P., and Troshina, V. P., 1968, "Changes in the sorption properties of glycerinated muscle models during their contraction," Tsitologiya, 10(12):1533.

Szent-Györgyi, A., 1941-1942, "The contraction of myosin threads," Stud. Inst. Med. Chem. Univ. Szeged, 1:17.

Szent-Györgyi, A., 1942, "The reversibility of the contraction of myosin threads," Stud. Inst. Med. Chem. Univ. Szeged, 2:25.

Szent-Györgyi, A., 1949, "Free energy relations and contractions of actomyosin," Biol. Bull., 96(2):140.

Szent-Györgyi, A., 1951, "The reversible depolymerisation of actin by potassium iodide," Arch. Biochem., 31(1):97.

Takata, M., 1961, "Studies on the protoplasmic streaming in the marine alga Acetabularia calyculus," Ann. Repts. Sci. Works, Fac. Sci. Osaka Univ., 9:63.

Tice, L. W., and Engel, A. G., 1966, "Cytochemistry of phosphatases of the sarcoplasmic reticulum. II. In situ localization of the Mg-dependent enzyme," J. Cell Biol., 31(3):489.

Townes, M. M., and Brown, D. E. S., 1965, "The involvement of pH, adenosine triphosphate, calcium, and magnesium in the contraction of the glycerinated stalks of Vorticella," J. Cell. Comp. Physiol., 65(2):261.

Ts'o, P. O. P., Bonner, I., Eggman, Z., and Vinograd, I., 1956a, "Observations on an ATP-sensitive protein from the plasmodium of a myxomycete," J. Gen. Physiol., 39(3):325.

Ts'o, P. O. P., Eggman, Z., and Vinograd, I., 1956b, "The isolation of myxomyosin, an ATP-sensitive protein from the plasmodium of a myxomycete," J. Gen. Physiol., 39(5):801.

Ts'o, P. O. P., Eggman, A., and Vinograd, I., 1957, "The interaction of myxomyosin with ATP," Arch. Biochem. Biophys., 66(1):64.

Ulbrecht, G., and Ulbrecht, M., 1952, "Der isolierte Arbeitszyklus glatter Muskulatur," Z. Naturforsch., 7b(8):434.

Ulbrecht, G., and Ulbrecht, M., 1957, "Phosphat-Austausch zwischen AMP und AD^{32}P durch hochgereinigte Aktomyosin-Präparate und gewaschene Muskelfibrillen," Biochim. Biophys. Acta, 25(1):100.

Ulbrecht, G., Ulbrecht, M., and Weber, A., 1954, "Die Beziehung zwischen Verkürzungsgeschwindigkeit und Belastung bei Fasermodellen," Biochim. Biophys. Acta, 13(4):564.

Ushakov, B.P., 1956, "The thermostability of cell proteins of cold-blooded animals in connection with species adaptation to environmental temperature conditions," Zh. Obshch. Biol., 27(2):154.

Ushakov, B.P., and Pashkova, I. M., 1967, "Intrapopulation variation in thermostability of muscles and glycerinated muscle fibers in various populations of the Black Sea mussel," in: Variation in Thermostability of Animal Cells During Ontogenesis and Phylogenesis [in Russian], Leningrad, pp. 74-81.

Ushakov, V. B., 1963, "The cause of thermal death of skeletal muscle fibers," Tsitologiya, 5(2):204.

Ushakov, V. B., 1964, "The cause of thermal death of skeletal muscles of cold-blooded animals," Dokl. Akad. Nauk SSSR, 155(5):1178.

Ushakov, V. B., 1965, "Thermostability of excitable and contractile systems of the muscle fiber," in: The Thermostability of Animal Cells [in Russian], Moscow—Leningrad, pp. 55-60.

Ushakov, V. B., 1966, "Comparative evaluation of the thermostability of muscles and myofibrils of Rana temporaria," Tsitologiya, 8(1):96.

Varga, L., 1946, "The relation of temperature and muscular contraction," Hung. Acta Physiol., 1(1):1.

Varga, L., 1950, "Observations on the glycerol-extracted musculus psoas of the rabbit," Enzymologia, 14(4):196.

Vol'fenzon, L. G., 1954, "Paranecrotic action of local anesthetics on cellular elements of various tissues," Zh. Obshch. Biol., 15(3):220.

Vorob'ev, V. I., and Ganelina, L. Sh., 1963, "Enzymic hydrolysis of ATP by glycerinated muscles during stretching," Tsitologiya, 5(6):672.

Vorob'ev, V. I., and Kukhareva, L. V., 1965, "Changes in adenosine triphosphatase activity of myosin during deformation in a hydrodynamic field," Dokl. Akad. Nauk SSSR, 165(2):435.

Vorob'eva, I. A., and Poglazov, B. F., 1963, "Isolation of a contractile protein from the alga Nitella flexilis," Biofizika, 8(4):427.

Watanabe, S., Hiroshige, T., and Tokura, S., 1960, "Isometric and isotonic studies on contraction and relaxation of glycerol-treated muscle fiber of rabbit psoas," J. Biochem. (Tokyo), 48(1):1.

Watson, M. R., and Hopkins, J. M., 1962, "Isolated cilia from Tetrahymena pyriformis," Exp. Cell Res., 28(2):280.

Weber, A., 1951, "Muskelkontraktion und Modelkontraktion," Biochim. Biophys. Acta, 7(1):214.

Weber, A., Herz, R., and Reiss, I., 1963, "On the mechanism of the relaxing effect of fragmented sarcoplasmic reticulum," J. Gen. Physiol., 46(4):679.

Weber, H. H., 1934, "Der Feinbau und die mechanischen Eigenschaften des Myosinfadens," Pflügers Arch. Ges. Physiol., 235:205.

Wilson, J. A., Elliott, P. R., Guthe, K. F., and Shappiro, D. G., 1959, "Oxygen uptake of glycerol extracted muscle fibres," Nature, 184:1947.

Winegrad, S., 1968, "Intracellular calcium movements of frog skeletal muscle during recovery from tetanus," J. Gen. Physiol., 51(1):65.

Winicur, S., 1967, "Reactivation of ethanol-calcium isolated cilia from Tetrahymena pyriformis," J. Cell Biol., 35(1):C7.